IMA MONOGRAPH SERIES

IMA MONOGRAPH SERIES

An Informal Introduction to Theoretical Fluid Mechanics

James Lighthill

Provost, University College London

CLARENDON PRESS · OXFORD

1986

Oxford University Press, Walton Street, Oxford OX2 6DP
Oxford New York Toronto
Delhi Bombay Calcutta Madras Karachi
Petaling Jaya Singapore Hong Kong Tokyo
Nairobi Dar es Salaam Cape Town
Melbourne Auckland
and associated companies in
Beirut Berlin Ibadan Nicosia

Oxford is a trade mark of Oxford University Press

Published in the United States
by Oxford University Press, New York

British Library Cataloguing in Publication Data

Lighthill, Sir James
An informal introduction to theoretical
fluid mechanics.—(IMA monograph series; 2)
1. Fluid mechanics
I. Title   II. Series
532   QC145.2
ISBN 0–19–853631–3
ISBN. 0–19–853630–5 Pbk

Library of Congress Cataloging in Publication Data

Lighthill, J., Sir.
An informal introduction to theoretical fluid
mechanics.
(IMA monograph series; 2)
Includes index.
1. Fluid dynamics. I. Title. II. Series.
QA911.L46 1986   532'.05   87–777
ISBN 0–19–853631–3
ISBN 0–19–853630–5 (pbk.)

Set by Macmillan India Ltd, Bangalore 25.
Printed in Great Britain by St Edmundsbury Press,
Bury St Edmunds, Suffolk

Preface

One of the major modern areas of successful practical application of mathematics is *fluid mechanics*. This is concerned with analysing the motion of either liquids or gases. Both of these, of course, are 'fluids'; that is, they are mobile substances lacking any large-scale order which are capable, consequently, of unlimited deformation and of yielding in time to any disturbing force however small.

Study of the mechanics of fluids is important in many contexts such as the following.

1 Locomotion through fluid media

All animals live immersed in fluid (air or water) and their capability of motion through it is of crucial importance for their life style. Man has greatly modified his life style by devising machines for improved locomotion. A most valuable flexibility is conferred when locomotion is achieved not by pushing the ground (as a walker or a train does) but by pushing the fluid (as in animal swimming or flying, or as in ships or aircraft). Study of all these matters (ranging from zoology to engineering) involves advanced fluid mechanics.

2 Circulation systems

Systems of circulating fluid offer important means for distributing things where they are needed. Once more, zoology gives a good illustration of this through the vital importance of the circulation of the blood. Modern chemical processing plant depends just as crucially upon the liquid or gaseous convection of dissolved chemicals, of suspended particles, or of energy in circulation systems. The atmosphere is a vast circulation system, driven by the heat of the sun, and engaged in the transport of heat, water-vapour, oxygen, carbon dioxide and various pollutants. The ocean is another great circulation system of practically equal importance to man; who needs to investigate the mechanics of the fluid motions involved in all of these.

3 Transfer of energy in engines

Energy stored as potential energy, chemical energy or heat energy becomes converted into kinetic energy in a water turbine, a gas turbine or a steam turbine, in each case by means of fluid flow acting on rotating blades. Such flow is studied in order to improve the efficiency of turbines; which may also, in many cases, depend upon effective fluid motions for transferring heat quickly from one part to another in such an engine.

4 Resistance of structures to wind and water

The design of structures intended to resist strong winds, river erosion, or violent sea motions requires an understanding of what determines the forces exerted by winds, currents or waves upon stationary structures. Although these are complex problems, we may note that two of the relatively less complex problems under headings 1 and 4 here are essentially the same: the fluid mechanics determining the resistance to a vehicle moving through still air; and the fluid mechanics determining the force of a steady wind on a stationary structure. It helps, indeed, to study the former problem *from the standpoint of the moving vehicle*: in a frame of reference in which the vehicle is at rest, the air is blowing past it with equal and opposite velocity, and the vehicle becomes effectively a stationary structure in a wind.

One of the reasons why the motions of fluids are so complex derives from the fundamental property that the fluid is capable of unlimited deformation (unlike solid structures which, in general, are capable of only a very limited degree of deformation without breaking). Some other reasons for complexity will be specially studied in this book under the headings of *boundary layers* and *turbulence*.

An essential characteristic of the application of mathematics to systems of great complexity (like fluids in motion) is that progress can be made only through an efficient cooperation between theory and experiment. There is a big contrast here with the study of much simpler systems for which the basic physical laws are known which allow computer programs to be developed that will predict reliably the behaviour of the system. Most fluid motions, as we shall see, are much too complex for that to be possible even if the largest and fastest of the nineteen-eighties generation of computers is being used.

Great progress with the effective study, and the effective computation, of fluid motions has been made, however, through the realization that such progress required creative inputs on a continuing basis both from theory and from experiment. Even though the basic physical laws underlying the mechanics of fluids are known with precision, typical problems encountered under headings 1 to 4 above involve motions of such complexity that the most powerful computers cannot infer those motions as a straightforward deductive exercise from those basic physical laws. At the same time, experiments on the intricate details of particular fluid motions are possible although they are in general very expensive. How, then, is it possible to make predictions about the vast majority of fluid motions: those which have not been subjected to such detailed experimental probing?

This book seeks to exemplify the answers to that question in an informal, introductory way. Briefly, those answers are based on the creative use of data from experimental studies and data from theoretical analyses to generate practically useful mathematical models (including manageable computer models) of a wide range of important fluid flows. Some of the analyses, as we shall see, involve mathematically exciting theories which, incidentally, are of a strikingly nonlinear character. The book's prime emphasis, however, is on the problem of how to use those as strong supports at one end of an effective bridge spanning the world of mathematics and the world of experiment and observation.

The illustrative examples given are concerned entirely with water and air, rather than with more complicated fluids such as blood or the fluids (including highly viscous fluids) used for various lubrication purposes. They do include something about circulation systems and about resistance, as well as about the fluid mechanics that both makes flight possible and underlies the energy transfer in certain types of turbine. Features special to flows at very high velocities (significantly greater than 100 m/s) or very low velocities (significantly less than 1 m/s) are left out, however, while for matters concerned with water waves, sound waves, shock waves and the mechanics of stratified fluids, readers are referred to the present author's *Waves in Fluids* (Cambridge University Press, 1978) which they may find a suitable text for study after they have read (say) Chapters 1 to 8 of the present work.

London   J.L.
July 1985

Contents

1
Principles of mechanics applied to lumps of fluid

All the analysis in this book is founded upon the basic principles of mechanics, such as Newton's laws of motion and the momentum and energy principles. These principles are assumed to be known, along with the most elementary physical properties of fluids. The present chapter, however, takes the study of fluid mechanics only as far as can be achieved by applying the principles of mechanics to 'lumps' of fluid on a large scale.[†]

1.1 Elementary mechanics of a fluid in equilibrium

The word **hydrostatics** is often used to describe the mechanics of a fluid in equilibrium; in other words, to describe the statics of a fluid. The elements of hydrostatics form a simple body of knowledge which is assumed to be known to the reader, and which is now briefly summarized.

The forces that act on a lump of fluid in equilibrium (Fig. 1) consist of (*i*) its weight and (*ii*) forces acting normal (that is, perpendicular) to its boundary.[‡]

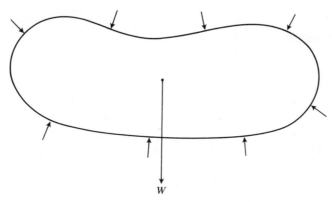

Fig. 1. The forces acting on a lump of fluid in equilibrium consist of its weight W and pressure forces acting normal to its boundary.

[†] Rather than by more systematically building up knowledge from their application to very small 'particles' of fluid as in later chapters.

[‡] Here is an important difference from a lump of solid material, which can be in equilibrium under a system of forces including tangential components.

The magnitude of these latter forces per unit area of boundary is called the pressure.

At each point in a fluid in equilibrium the pressure has a definite value which for *any* lump of fluid with that point on its boundary is the normal force per unit area acting on that lump. In other words, the pressures acting in every direction are equal.

Another very well known result, that the pressure is the same at all points of the fluid which are at the same height, has some quite remarkable consequences. It means that a lump of fluid can be used (Fig. 2) to permit a kilogram weight to balance a tonne if their areas of application are in the ratio 1 : 1000. This is the principle of the hydraulic press: a device which can have an almost indefinitely large 'mechanical advantage' (ratio of the output force, used here to support, or perhaps to raise, the tonne weight, to the input force, here supplied by a weight of one kilogram; or, for raising, slightly more).

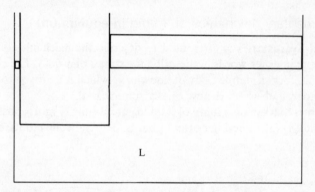

L

Fig. 2.   Schematic illustration of the principle of the hydraulic press: the small weight acting on a small area generates the same fluid pressure as the much larger weight acting on a proportionately larger area.

The balance of forces upon a lump of fluid in equilibrium (Fig. 1) tells us that the *resultant* of the normal pressures must be a force equal and opposite to the weight of the lump. This is Archimedes' Principle, which for fluid of effectively uniform density ρ (that is, with negligible *stratification* of density), gives the upward resultant as

$$\rho Vg, \tag{1}$$

where V is the volume of the lump and g is the acceleration due to gravity. This result applies also if the lump is replaced by a solid lump of the same size and shape: the distribution of pressures in the fluid is unchanged and they have the same upward resultant ρVg. In words, the *buoyancy force* on the body is equal to the weight of fluid which the body has displaced.

Finally, Archimedes' Principle applied to a lump in the form of a cylinder with vertical generators, which stretches between two levels at heights H_1 and H_2 above some 'reference level' (e.g. the ground), tells us that the difference in the pressures p_1 and p_2 at the two levels is

$$p_2 - p_1 = \rho g(H_1 - H_2);$$ (2)

an equation identifying buoyancy force with weight per unit cross-sectional area of that cylinder. The distribution of pressure in a fluid of uniform density in equilibrium is, in short, specified by the rule

$$p + \rho g H = \text{constant.}$$ (3)

1.2 Flow through a contraction in a horizontal pipe

From elementary statics we now move to how the basic principles of dynamics can be applied to fluids in motion. The impact of a horizontal jet of fluid, of density ρ, velocity v, and cross-sectional area S, on a wall at right angles to the jet is commonly used in dynamics texts to illustrate the application of the momentum principle. The jet delivers a mass of fluid $\rho S v$ per second, and so the jet force on the wall is estimated as

$$\rho S v^2,$$ (4)

this being the rate of delivery by the jet of horizontal momentum, all of which is assumed destroyed at the wall.

Even if we are satisfied with relatively crude estimates, however, there are severe limitations to the range of problems for which such simple consider-ations of mass and momentum can give useful information. In order to illustrate this we study the flow of fluid through a gradual contraction in a horizontal pipe (Fig. 3), and focus attention on a large lump L of fluid

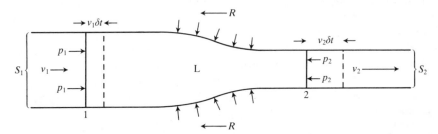

Fig. 3. Flow through a horizontal pipe with a gradual contraction in cross-sectional area from S_1 to S_2. In time δt the lump L of fluid moves from a position between the two plain lines at stations 1 and 2 to a position between the two broken lines. Pressure forces, shown as acting normally to the pipe wall in the region of the contraction, have the horizontal resultant R which opposes the net pressure force $p_1 S_1 - p_2 S_2$ acting on the lump L at stations 1 and 2.

stretching from station 1, of cross-sectional area S_1, upstream of the contraction, to station 2, of smaller cross-sectional area S_2, downstream of it.

In a small time δt, the front face of lump L is carried downstream a distance $v_2 \delta t$ where v_2 is the fluid velocity at station 2, and the back face is carried downstream a distance $v_1 \delta t$ where v_1 is the velocity at station 1. Consequently, the region occupied by lump L changes, by taking in an additional volume $S_2 v_2 \delta t$ at the front and vacating a region of volume $S_1 v_1 \delta t$ at the back. The *rate of change* of mass for lump L consists then of two terms: from the additional volume a positive contribution $\rho S_2 v_2$ and from the vacated volume a negative contribution $-\rho S_1 v_1$. As the mass is necessarily conserved we deduce that at each instant

$$S_1 v_1 = S_2 v_2, \tag{5}$$

where both sides of the equation represent the instantaneous rate of volume flow through the pipe.

We now give particular consideration to a 'steady-flow' case, i.e. one in which the rate of volume flow is not changing with time. Although the volume flow is constant, eqn (5) shows that $v_2 > v_1$ (as $S_2 < S_1$); the fluid *speeds up* as it enters the contraction.

What force produces this acceleration? In order to generate such a force it seems clear that the fluid pressure p_1 at station 1 must exceed the pressure p_2 at station 2, but the amount of that excess is not easily estimated. For example, consideration of the rate of change of momentum for lump L fails to determine the pressure drop $p_1 - p_2$ because the *horizontal resultant R* of the pressure forces between the pipe and the fluid is unknown (see Fig. 3 where these forces act at right angles to the boundary so that in the region of the contraction they must possess components along the axis of the pipe).

The net horizontal force accelerating lump L is $p_1 S_1 - p_2 S_2 - R$, consisting of a pressure force at station 1 opposed by a smaller pressure force at station 2 and by the reaction at the pipe. The corresponding rate of change of momentum consists of a positive contribution $\rho S_2 v_2^2$ at station 2 (the velocity v_2 times the rate of change of mass $\rho S_2 v_2$ due to lump L moving into an additional volume at the front) and a negative contribution $-\rho S_1 v_1^2$ at station 1 (similarly associated with the volume vacated by lump L at the back; note that in steady flow the momentum in any fixed region of space remains constant, so that any change in the momentum of a lump of fluid arises from changes in the region it occupies). Equating the force to the rate of change of momentum gives us an equation

$$p_1 S_1 - p_2 S_2 - R = \rho S_2 v_2^2 - \rho S_1 v_1^2, \tag{6}$$

but as this equation (to which we return in Section 1.4) includes the unknown reaction R it fails to give us information about the pressure drop $p_1 - p_2$.

The way to obtain a useful estimate of that pressure drop is, rather, by considerations of energy. The rate of change of kinetic energy of lump L consists of a positive contribution $(\frac{1}{2}v_2^2)(\rho S_2 v_2)$ at station 2 (half the velocity squared times the rate of change of mass) and a corresponding negative contribution $-(\frac{1}{2}v_1^2)(\rho S_1 v_1)$ at station 1. If we equate this rate of change to the rate at which external forces do work on the lump L we obtain a positive contribution $(p_1 S_1)v_1$ (force times velocity component in the direction of the force) at station 1, while the contribution $-(p_2 S_2)v_2$ from station 2 is negative (because the force is in the opposite direction to the velocity). On the other hand, the pressure forces between the pipe wall and the fluid act at right angles to the velocity (Fig. 3) *and therefore can do no work*, so that, on this analysis, the unknown reaction R makes no contribution.

Equating the rate at which work is done on lump L to its rate of change of energy, we obtain the equation

$$p_1 S_1 v_1 - p_2 S_2 v_2 = (\tfrac{1}{2}v_2^2)(\rho S_2 v_2) - (\tfrac{1}{2}v_1^2)(\rho S_1 v_1) \tag{7}$$

which can be simplified by taking out a factor given by either side of eqn (5) to give

$$p_1 - p_2 = \tfrac{1}{2}\rho v_2^2 - \tfrac{1}{2}\rho v_1^2. \tag{8}$$

This equation for estimating the pressure drop is discussed further, and critically examined, in the next section.

1.3 The total head of a steady stream

There are several reasons why eqn (8), although often very useful, can at most be only a rough approximation to the pressure drop which, indeed, it tends to underestimate. The first of these reasons is very familiar from elementary mechanics, where energy arguments may produce only crudely approximate results when they neglect the dissipation of kinetic energy due to *friction.*

Indeed, frictional forces (that is tangential forces) do occur between a solid boundary and a *moving* fluid, even though the well established and experimentally well supported laws of hydrostatics (Section 1.1) rule out such forces for a fluid at rest. Later, we shall see that the magnitude of any tangential force acting at the boundary of a lump of fluid that moves in a particular flow pattern depends on a well defined physical property of the fluid called the *viscosity.* We shall find, furthermore, that the fluids upon which this book concentrates (air and water) are fluids of *small* viscosity, in a certain well defined sense.

It might be tempting to infer in rather general terms from this that energy arguments such as were used to derive eqn (8) give results which are good approximations for air and water. The truth is more complicated, however. Equation (8) gives results that agree quite reasonably with experiment for the case actually illustrated in Fig. 3 (a gradual contraction) but gives results that are very inaccurate for some other pipe geometries, for example an *expansion* of cross-section (see Section 1.5), or an *abrupt* contraction.

Much of the present book is concerned with expounding the intricate reasons for these distinctions. They are important not merely because they set limits on the applicability of a method of calculation but also because (as in other branches of mechanics) the motions which keep energy dissipation to a minimum are most advantageous for many engineering applications. In this section, however, we just briefly initiate that discussion.

Frictional forces can be important not only between a lump of fluid and its *solid* boundary (as so far discussed) but also between neighbouring lumps of fluid. Such *internal* friction is able to dissipate kinetic energy into heat energy. Admittedly, the rate of frictional dissipation for a fluid moving in a particular flow pattern is another quantity proportional to the viscosity of the fluid. Nevertheless, there are certain pipe geometries which (for reasons that will emerge later) lead to flow patterns of a type especially prone to dissipate energy even for fluids of very small viscosities, and this ruins the accuracy of eqn (8).

One further obstacle to the accuracy of the equations of Section 1.2 exists. Friction tends to produce an uneven distribution of fluid velocities across the pipe, with the flow retarded more near a solid wall; yet in the arguments leading to eqn (5), for example, the velocity v_1 was assumed uniform across the cross-section (and similarly with v_2). Admittedly, a detailed study of those arguments shows that eqn (5) must remain correct if v_1 is the fluid velocity at station 1 *averaged* across the cross-sectional area (and similarly with v_2). However, this interpretation makes for difficulties in eqn (6) which, on a similar basis, would remain correct only if v_1^2 were the *fluid velocity squared* averaged over the cross-sectional area. Clearly, this is incompatible with the former determination of v_1 since the average of the square of a quantity always exceeds the square of its average. Similar difficulties arise in eqns (7) and (8).

For certain pipe geometries, however, including that of Fig. 3, typical flows of water (or of air) in the ranges of speed studied in this book involve velocity distributions that are almost uniform across the pipe except *very* near the wall. In such a case the above difficulties (as well as those others remarked on earlier) lead to only modest errors.

If the pipe in Fig. 3 is not horizontal, stations 1 and 2 may be at different heights H_1 and H_2 above the ground. In this case, the rate of change of potential energy due to gravity (gH per unit mass) has to be added on to the right-hand side of eqn (7), giving an extra term

$$(gH_2)(\rho S_2 v_2) - (gH_1)(\rho S_1 v_1). \tag{9}$$

When the factor $S_2 v_2$ (or $S_1 v_1$, which is the same) is taken out of expression (9), this expression provides an extra term $\rho g H_2 - \rho g H_1$ on the right-hand side of eqn (8), which can then be written

$$p_1 + \rho g H_1 + \tfrac{1}{2}\rho v_1^2 = p_2 + \rho g H_2 + \tfrac{1}{2}\rho v_2^2. \tag{10}$$

The quantity

$$p + \rho g H + \tfrac{1}{2}\rho v^2, \tag{11}$$

which eqn (10) specifies as taking identical values at stations 1 and 2 (in steady-flow conditions with negligible energy dissipation), is usually called the *total head* of a stream of velocity v. The first two terms (total head when $v = 0$) give a result completely compatible with the hydrostatic rule (3). The last term represents an addition sometimes known as the *dynamic head*. The *last two* terms represent the mechanical energy (potential plus kinetic) per unit volume of fluid, but it is the whole expression (including the first term) which must be used in the case of a steady stream of fluid when we express the absence of energy dissipation.

In practice, of course, there is always some energy dissipation, which means that the total head downstream (right-hand side of eqn (10)) is necessarily *less* than its value upstream (the left-hand side). Thus eqns (8) or (10) underestimate the actual pressure drop. However, in those many engineering circumstances where dissipation of energy is disadvantageous, engineers are concerned to design systems so that they keep to a minimum this loss of total head. Designers must, in short, be miserly of total head.

As just one simple illustration of the above remarks we consider a wide tank discharging in a narrow horizontal jet (Fig. 4). Choosing station 2 in the jet, just outside the orifice, we take $H_2 = 0$ (with the reference level chosen as that of the orifice). Station 1 is taken as the tank's free surface, which is supposed to be of area S_1 so much greater than the jet cross-sectional area S_2 that eqn (5) makes v_1 negligible compared with v_2.

Actually, a design like that of Fig. 4 is appropriate if loss of total head is to be

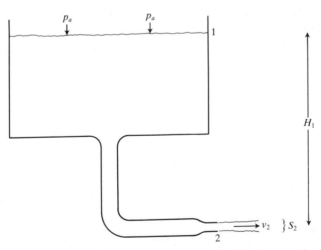

Fig. 4. A wide tank is discharging water in a narrow horizontal jet of cross-sectional area S_2 at a vertical distance H_1 below the free water surface.

kept *small* in order to maximize the velocity v_2 of the jet for a given height H_1 of the free surface above the orifice. Then eqn (10) gives

$$p_a + \rho g H_1 + 0 = p_a + 0 + \tfrac{1}{2}\rho v_2^2, \quad \text{or} \quad v_2 = (2gH_1)^{\frac{1}{2}}, \tag{12}$$

where the pressures p_1 and p_2 are both equal to the pressure p_a of the atmosphere with which the water is in contact at both stations.

The *smoothness* of the shapes in Fig. 4, both of the contractions (at the outflow from the tank and just before the jet orifice) and also of the pipe's right-angle bend, helps to reduce loss of total head; while the very existence of the latter contraction means that energy dissipation in the main part of the pipe is kept down because the local velocity is far less than v_2. Any conditions substantially different from these would bring about a significant loss of total head, usually found to take the form $\tfrac{1}{2}\rho v_2^2 k$, where k depends on the pipe geometry. The eqns (12) are then replaced by

$$\rho g H_1 = \tfrac{1}{2}\rho v_2^2 (1+k), \quad \text{or} \quad v_2 = (2gH_1)^{\frac{1}{2}}(1+k)^{-\frac{1}{2}}, \tag{13}$$

showing the reduction in jet speed associated with loss of total head.

1.4 Reaction forces on pipes carrying flows

Having analysed the pressure drop in a contraction under conditions favourable to small energy dissipation, we may go back to eqn (6), which we derived from the momentum principle, and which we are now able to use to determine the reaction of the pipe

$$R = (p_1 + \rho v_1^2)S_1 - (p_2 + \rho v_2^2)S_2. \tag{14}$$

Besides being the force with which the solid wall of the contraction acts on the water flowing through it, we know by Newton's third law that the water acts with an equal and opposite force R on the pipe. This force is communicated to the structure, or the person (e.g. a fireman), that may be holding the pipe in place.

Substituting for p_1 from eqn (8) in the eqn (14) for R, we obtain

$$R = (p_2 + \tfrac{1}{2}\rho v_2^2 + \tfrac{1}{2}\rho v_1^2)S_1 - (p_2 + \rho v_2^2)S_2; \tag{15}$$

and, by using eqn (5) to replace v_1 by $v_2 S_2/S_1$, we obtain

$$R = p_2(S_1 - S_2) + \tfrac{1}{2}\rho v_2^2 (S_1 - S_2)^2/S_1. \tag{16}$$

This is the force required to hold in place the pipe of Fig. 3.

Where the stream on the right-hand side discharges into the atmosphere, the pressure p_2 in eqn (16) takes the atmospheric value p_a, but the term $p_a(S_1 - S_2)$ is exactly balanced out by the action of atmospheric pressure on the outside of the pipe. Therefore, it is only the last term in eqn (16) which represents the force actually needed to be exerted by a fireman to hold the nozzle of a hose. Note that, for reasons mentioned at the end of Section 1.3, the loss of total head in the hose as a whole can be reduced by the choice of a sizeable contraction ratio S_1/S_2, but that there are limits as to how far S_1 can be increased for a given jet velocity v_2 because the force needed to hold the nozzle (given by the last term in eqn (16)) may become too large.

Another interesting form of R is obtained when we use eqns (8) and (5) combined in the form

$$p_1 - p_2 = \tfrac{1}{2}\rho v_2^2 (S_1^2 - S_2^2)/S_1^2 \tag{17}$$

in order to eliminate $\tfrac{1}{2}\rho v_2^2$ from eqn (16), giving

$$R = p_2(S_1 - S_2) + (p_1 - p_2)\left(\frac{S_1(S_1 - S_2)}{S_1 + S_2}\right) = \left[\frac{p_1 S_1 + p_2 S_2}{S_1 + S_2}\right](S_1 - S_2). \tag{18}$$

Equation (18) represents the resultant force that would be exerted if a *uniform* pressure equal to the expression in square brackets acted over the whole surface of the contraction. In a certain sense, then, this expression in square brackets represents an *average* pressure acting in the contraction region to produce the resultant force R. As might be expected, this value of the pressure takes a value intermediate between p_1 and p_2.

Arguments from the momentum principle can also be used to determine the reaction force in curved contraction regions of pipes, provided that as in Fig. 5 they are smoothly shaped so that the loss of total head can be assumed to be small. Those arguments tell us, for example, that the force with which the water acts on the pipe in Fig. 5 would be the resultant of the three forces shown. In the

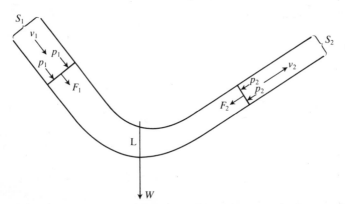

Fig. 5. The force with which a lump L of fluid flowing through a curved contraction in a pipe acts upon it is the resultant of: (i) a force $F_1 = p_1 S_1 + \rho S_1 V_1^2$ directed as shown; (ii) another force $F_2 = p_2 S_2 + \rho S_2 V_2^2$ directed as shown; and (iii) the weight W of lump L.

absence of flow they would be the pressure force $p_1 S_1$ at station 1, the pressure force $p_2 S_2$ at station 2 and the weight of the fluid. In steady flow these values are augmented by two terms representing rate of change of momentum ($\rho S_1 v_1^2$ at station 1 and $\rho S_2 v_2^2$ at station 2), and a careful statement of the momentum principle for the fluid readily equates the force on the pipe (being equal and opposite to the reaction of the pipe on the fluid) to the vector sum of the forces in Fig. 5, with these terms directed as shown.[†]

1.5 Loss of total head at an abrupt expansion

Although there is relatively little loss of total head in a gradual contraction, the situation is quite different in a region of a pipe where its cross-section increases (an expansion region). Where a fluid flow enters a gradual expansion region there is (for reasons to be gone into later) a strong tendency for the main fluid flow to become 'separated' from the pipe wall: the main part of the flow fails to undergo the full expansion in its diameter which the pipe undergoes (see Fig. 6). This not only accentuates the problems of uneven distribution of velocity in the pipe (with the flow much more retarded near the wall) mentioned in Section 1.3, but also leads to greatly increased dissipation of energy in an internal region of particularly intense chaotic motion or 'turbulence'.

Engineers are, however, able with very careful design to construct 'diffusers': extremely gradual expansion regions shaped so as to avoid separation and to keep the loss of total head small (Fig. 7). In this case the equations of Section 1.2

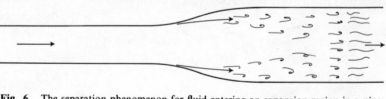

Fig. 6.   The separation phenomenon for fluid entering an expansion region in a pipe.

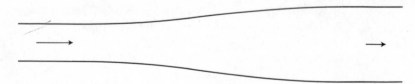

Fig. 7.   Separation may be avoided in an extremely gradual expansion region or 'diffuser'.

[†] Here, too, since the atmospheric pressure p_a acts all round the outside of the pipe, the total force on the pipe itself is best expressed as above with only the excess of the pressures p_1 and p_2 over p_a included.

apply; in particular, as the cross-sectional area increases (with, from eqn (5), a *reduction* in fluid velocity) there is, from eqn (8), a substantial recovery of pressure which may be of great importance in certain engineering systems.

The loss of total head in an expansion region may vary, then, from the low values characteristic of a well designed diffuser shape through a range of considerably larger values characteristic of smooth expansions in general. The greatest possible loss of head is found in a completely abrupt expansion (Fig. 8). In this case a jet is formed at the lip of the expansion; and a particularly intense turbulence at the edge of that jet dissipates a substantial amount of energy and also helps to spread the jet until, ultimately, it becomes a stream filling the whole expanded section of the pipe.

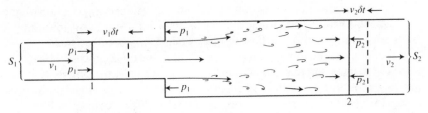

Fig. 8. In flow through an abrupt expansion, the main flow separates to form a turbulent jet surrounded by an almost stagnant *annulus* region.

This extreme case of an abrupt expansion is one in which, exceptionally, it is possible to use the momentum principle to calculate the pressure rise and, therefore, to calculate how much total head is lost. The reason for this is that the horizontal resultant of the pressure forces between the pipe and the fluid can be independently estimated with good accuracy.

In the *annulus* region of the solid pipe wall, where its area expands from S_1 to S_2, the pressures can indeed be estimated to be p_1, the pressure at station 1. This is an almost stagnant region of fluid because the main flow separates completely at the lip. The pressure in that stagnant region adjacent to the jet must be closely equal to that in the jet itself, and this, in turn, must closely equal the pressure p_1 immediately upstream of the lip since the jet has as yet undergone no change of cross-sectional area to alter its velocity and hence its pressure.

All this means that the resultant pressure force on the annulus region (of area $S_2 - S_1$) is $p_1(S_2 - S_1)$, so that the application of the momentum principle as in eqn (6) gives

$$p_1 S_1 + p_1 (S_2 - S_1) - p_2 S_2 = \rho S_2 v_2^2 - \rho S_1 v_1^2, \tag{19}$$

where the left-hand side is the total force acting on the lump L of fluid and the right-hand side is its rate of addition of momentum at station 2 minus its rate of loss of momentum at station 1. Equation (19), with the value of v_2 substituted from eqn (5), gives the pressure rise as

$$p_2 - p_1 = \rho v_1^2 [(S_1/S_2) - (S_1/S_2)^2]. \tag{20}$$

If this is plotted against the area ratio S_1/S_2 we see that the largest possible pressure rise (attained when $S_1/S_2 = \frac{1}{2}$) is $\frac{1}{4}\rho v_1^2$, only half what could be achieved in an ideal diffuser shape.

The loss of total head

$$(p_1 + \tfrac{1}{2}\rho v_1^2) - (p_2 + \tfrac{1}{2}\rho v_2^2) \tag{21}$$

can be deduced from eqn (20) in the form

$$\tfrac{1}{2}\rho v_1^2 [1 - (S_1/S_2)]^2. \tag{22}$$

This is seen to be a modest fraction of the initial kinetic energy per unit volume (or 'dynamic head') $\frac{1}{2}\rho v_1^2$ when S_1/S_2 is only a little less than unity, but it becomes a rather larger fraction of $\frac{1}{2}\rho v_1^2$ when S_1/S_2 becomes small.

If we compare the forces acting on the fluid in the different cases (Figs. 6, 7, and 8) we can see that this loss of total head in an abrupt expansion represents a *maximum* loss for an expansion from area S_1 to area S_2. The ideal diffuser (Fig. 7) achieves the greatest possible pressure rise $p_2 - p_1$, corresponding to no loss of total head, and the average pressure in the expansion region then takes the value given in square brackets in eqn (18). This is a value substantially larger than p_1 and, when it replaces p_1 in the second term of the momentum equation (eqn (19)), it allows p_2 to take its maximum value. The other extreme is the case of Fig. 8, where the horizontal resultant of pressure forces between the pipe and the fluid involves *no increase at all* above the pressure of the oncoming stream, leading to the greatly reduced pressure rise of eqn (20). Intermediate values are realised in intermediate cases like that of Fig. 6.

In Chapter 1 we have looked at a modest range of confined flows in which it is possible to derive a few results of value by applying the principles of mechanics to lumps of fluid on a large scale. It must be clear, however, that the methods used are very limited in their applicability, especially because of their concentration on just one degree of freedom (the velocity of the stream, averaged if necessary across the pipe). In order to make further progress in the mechanics of fluids it is essential to adopt some more refined, and more fully three-dimensional, methods of analysis.

2
Velocity fields and pressure fields

Fundamental mechanics makes most fruitful use of the concept of a 'particle': an element of matter of negligible dimensions. The mechanics of matter on a larger scale is often analysed by:

(i) regarding it as made up of very many small particles;
(ii) applying mechanical principles to each particle; and
(iii) using calculus to integrate up for the whole system.

Ordinary one-dimensional calculus suffices when this programme is applied to obtain, say, the shape of a hanging chain. It was Leonhard Euler (1707–83) who originated the three-dimensional calculus of vector fields (as we now call them) which is necessary in order to apply the same ideas to obtain a mathematical model of the mechanics of a fluid. The present chapter describes, and critically reviews, the foundations of the Euler model.[†]

2.1 Critical description of a mathematical model

A typical 'particle' of fluid, to which principles of mechanics are applied in mathematical models such as that of Euler, is illustrated in Fig. 9. The 'dimensions' of the particle (which we could call its 'length' dx, 'breadth' dy, and 'height' dz) are supposed negligible in the sense that they are very much smaller than any dimensions relevant to the system as a whole. The Euler model differs from that of Section 1.2 primarily by being founded on the application of principles of mechanics to a particle as in Fig. 9 instead of to a

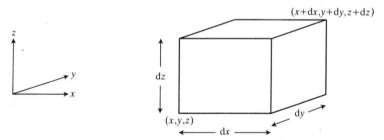

Fig. 9. A box-shaped particle of fluid, of length dx, breadth dy and height dz.

[†] Although this usage can without doubt be justified as referring to the model mainly used by Euler, our phrase 'the Euler model' is really used just as a convenient name for the model defined in Section 2.1.

lump L of fluid on a large scale as in Fig. 3, while in many other respects the approaches are very similar.

Thus, we shall see that conservation of mass for the particle of fluid is expressed by requiring that the rates at which its movement brings it to occupy additional volume and to vacate previously occupied volume are identical. The net pressure force on the particle is calculated from the differences in the opposing pressure forces acting on opposite faces; and this force, combined with that of gravity, is equated to the particle's rate of change of momentum. A new feature of both calculations, however, is that because we are concerned with differences in the values of quantities on the two sides of a particle of extremely small dimensions, those differences can be expressed in terms of differential coefficients, leading (as we shall see) to statements of mechanical principles in the form of *differential equations*. Before proceeding with this programme, however, we carefully list and discuss the three key assumptions underlying the Euler model.

First, the model regards the fluid as *continuous*, in the sense that the fluid properties measured on the scale of a large lump of fluid are assumed to be fully reproduced on the scale of an extremely small particle as illustrated in Fig. 9. This assumption does need careful questioning because of our intention to replace differences in the values of quantities on both sides of the particle in terms of differential coefficients. We shall wish, for example, to write the difference between the pressure at the point (x, y, z) and that at the point $(x + dx, y, z)$ as

$$(\partial p / \partial x)\,dx. \tag{23}$$

Yet the definition of the partial differential coefficient $\partial p / \partial x$ makes it the *limiting value* of the ratio of the pressure difference to dx as the particle's dimension dx *tends to zero*. This seems to imply a requirement that the particle's dimension dx must be capable of being reduced without limit to the point where the assumption of continuity would conflict with the real molecular structure of fluids.

Careful reconsideration of this problem shows, however, that the assumption leads to no significant inaccuracy in the model as applied to fluids like air and water. To validate this claim we require only that eqn (23) gives the difference in pressure to an extremely close approximation. The dimension dx must be small enough, then, for the variation in the pressure p across the particle to be effectively linear, making the above pressure difference linearly proportional to dx, as in eqn (23), without any higher terms in its Taylor series becoming significant.[†]

[†] In this book, as is frequently found convenient in applied mathematics, we avoid the use of δx for a small but finite difference between two values of a variable x; instead, we use dx provided that the difference in question is small enough for the dependent variables (like pressure and velocity) with which we are concerned to show a closely linear variation across a distance dx so that expressions like eqn (23) can be used with very good accuracy.

The particle dimensions must be small, then, in the above sense, meaning essentially that, as stated at the beginning of this section, they are very much smaller than any dimensions relevant to the fluid motion as a whole. Fortunately, it is readily possible for particles small in *this* sense to be extremely large in the context of the molecular structure of the fluid! This condition only requires such particles, in a liquid, to be large compared with the size of a molecule (under 10^{-6} mm for water) or, in a gas, to be large compared with the mean free path between molecular collisions (around 10^{-4} mm for air at atmospheric pressure). If we exclude exceptional fluids, then, such as solutions of extremely large molecules, or highly rarefied gases, we can fully substantiate the *continuity* assumption in the Euler model.[†]

The second of the model's key assumptions is that the fluid is *incompressible*. The model proceeds, in fact, as in Section 1.3, by treating the density ρ as constant. This assumption, of course, can never be exact because any increase (for example) in the fluid pressure produces some corresponding increase in density. We can, however, provisionally identify a wide class of motions of fluids for which any density changes must be a negligibly small fraction of the density.

That fractional change in density is given, in fact, by $K(p_2 - p_1)$, where K is the fluid's compressibility, when the pressure changes from p_1 to p_2. Results of the general form of eqn (8) (which will be found in Section 3.5 to appear also in the Euler model) indicate that pressure changes of the magnitude of $\frac{1}{2}\rho v^2$ accompany flows of typical velocity v. The fractional change in density is very small, then, if $\frac{1}{2}K\rho v^2$ is very small. For water this quantity is less than 0.01 (corresponding to a density change of less than one per cent) for $v < 200$ m/s, a very large upper limit which means that all practical water flows can be considered incompressible. Similar conclusions are reached for other liquids.

For air, however, the corresponding upper limit on velocity which is required for density changes to be under one per cent is 50 m/s. This means that very many airflows of practical interest, including all those at the speeds of natural winds, can be considered incompressible, but that many high-speed flows of air and other gases that are important for engineering applications do not satisfy this condition. For these flows a modified model taking density variations into account (along lines very briefly indicated in Section 3.1) is necessary. Actually, more refined studies using this modified model indicate that the Euler model gives reasonably good results for airspeeds v up to about 100 m/s (the upper limit to the scope of this book which was given in the Preface) but that it should not be used at higher speeds.

[†] We should, however, note one special property of liquids, including water, which may occasionally need to be taken into account as disturbing the *continuity* assumption in the Euler model. This is their tendency to 'cavitate' (to form an assemblage of small bubbles) if the pressure actually reaches *negative* values; for example, if the pressure drop given by equation (8) actually exceeds the initial pressure p_1. Engineering design usually seeks to avoid cavitation.

The third, and by far the most limiting, of the model's key assumptions is that the fluid is *frictionless*. This assumption, discussed previously in Section 1.3, neglects the fact that for a fluid *in motion* tangential forces (and not merely normal pressure forces) may act between adjacent particles if they are moving at different velocities. The coefficient in the relationship between tangential force per unit area and velocity *gradient* is called the viscosity μ. Although this takes numerically small values (of the order $10^{-5}\,\mathrm{N\,s\,m^{-2}}$ for air and $10^{-3}\,\mathrm{N\,s\,m^{-2}}$ for water) we shall find that there are special reasons why in many flows such tangential forces produce very important effects (often connected, as in Section 1.5, with flow *separation* from a solid surface). By contrast, they are unimportant for many other flows, including flows past bodies of so-called 'streamlined' shape, and we shall find that there are often major engineering advantages in designs which allow a flow close to that of the frictionless Euler model to be achieved.

The most serious limitation on the accuracy of the Euler model is, then, its assumption of frictionless flow. Nevertheless, it is valuable to develop the model in detail (especially because the achievement of flows close to those it predicts may in many contexts be desirable) provided that we simultaneously acquire insight into the range of conditions under which the model's predictions are borne out in practice. Parts of every subsequent chapter are devoted to the objective of giving such insight; but the remainder of the present chapter is concerned with laying the foundations for the Euler model itself.

2.2 The pressure field and its gradient

Once we assume that, even for a fluid in motion, the surrounding fluid acts on a particle of that fluid by means of forces which are purely *normal* pressure forces, the demonstration that the pressure acting in every direction must be the same is straightforward. The following consideration of a small particle in the shape of a right-angled triangular prism (Fig. 10), where the angle θ is arbitrary, shows that the pressures p_1 and p_2 become identical as the size of the prism tends to zero. In fact, the pressure p_1 acts on an area $hl \cos \theta$ (where h is the height of the prism perpendicular to the paper) and the pressure p_2 acts on an area hl. The net pressure force on the prism resolved in the direction of the pressure force p_1 is therefore

$$p_1 \,(hl \cos \theta) - (p_2 \, hl) \cos \theta. \tag{24}$$

But the mass of the prism is

$$\rho \left(\tfrac{1}{2} hl^2 \sin \theta \cos \theta \right) \tag{25}$$

and Newton's second law equates the force in expression (24) to the mass in expression (25) multiplied by the *difference* between the acceleration of the particle and the acceleration due to gravity, both resolved in the direction of

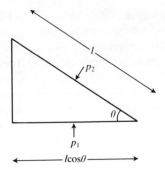

Fig. 10. A particle of fluid in the shape of a right-angled prism with dimensions h (perpendicular to the paper) and l (length of sloping face).

the pressure force p_1. In this equation the term which includes the mass thus includes the factor l^2 and therefore disappears in the limit as $l \to 0$, when the equation gives simply $p_1 = p_2$.

In the Euler model, then, the spatial distribution of a single quantity, the fluid pressure p, determines the internal force acting upon a particle of fluid. Often, it is convenient to think geometrically about that distribution of pressure in terms of the shapes of the surfaces $p = $ constant (the 'isobars'). The assemblage of all these different constant-pressure surfaces is commonly called the *pressure field* (Fig. 11). Although the extremely simple pressure distribution (eqn (3)) for a fluid at rest implies that these constant-pressure surfaces are simply parallel horizontal planes, fluids in motion have pressure fields of more complicated shapes (Fig. 11) such as are familiar from weather maps.

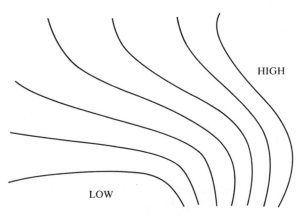

Fig. 11. A pressure field of isobaric surfaces $p = $ constant.

The geometry of the pressure field helps to give a clear picture of both the magnitude and direction of a very important vector: the *pressure force per unit volume* on a particle of fluid. Figure 12 shows two very closely neighbouring isobaric surfaces on one of which the pressure takes the constant value p, while on the other it takes the *very* slightly larger value $p + \mathrm{d}p$ (see footnote to section 2.1 for this use of d). Locally, the surfaces are separated by a distance $\mathrm{d}n$, measured normally to them. The figure shows a particle of fluid in the form of a cylinder of very small height $\mathrm{d}n$ and very small base area S.

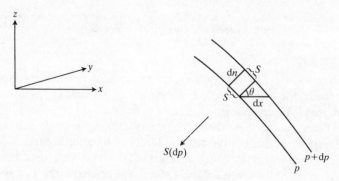

Fig. 12.   The force on a cylindrical particle of fluid of cross-section S, situated between two isobaric surfaces a small distance $\mathrm{d}n$ apart.

The net pressure force on the cylindrical particle in Fig. 12 is seen to have the value

$$S(p+\mathrm{d}p) - Sp \text{ in the direction } p \text{ decreasing.} \tag{26}$$

Note that there are only these two contributions to the net pressure force (from the top face and from the bottom face of the cylinder, respectively). In fact, at each height *between* the top and the bottom the pressures are the same all round the cylinder's cross-section and therefore can exert no net force.

The volume of the cylinder is the base area S times the height $\mathrm{d}n$, so that the pressure force per unit volume on the particle is $\mathrm{d}p/\mathrm{d}n$, acting once more in the direction p decreasing (a direction sometimes described as 'down the gradient' of pressure). For a small particle in the sense described in Section 2.1, this ratio can be written as a partial differential coefficient

$$\partial p/\partial n, \tag{27}$$

meaning the rate of change of p with distance along the normal, this time acting in the direction p increasing ('up the gradient' of pressure).

The word gradient is used twice in the previous paragraph in the ordinary sense familiar from topographical contour maps where 'up the gradient' indicates a direction of steepest ascent, normal to the contours, and 'down the gradient' one of steepest descent. In the analysis of pressure fields (and other three-dimensional scalar fields) the word gradient is, however, given a very precise mathematical sense as a quantity with magnitude as well as direction: the vector grad p (also written ∇p) is defined as a vector of magnitude

$$\partial p/\partial n \text{ in the direction } p \text{ increasing.} \tag{28}$$

According to this standard definition

$$\text{pressure force per unit volume} = -\text{grad } p, \tag{29}$$

since that force per unit volume is established to be of magnitude $\partial p/\partial n$ in the direction p *decreasing* by the argument given between eqns (26) and (27).

We often need to know the component of grad p along one of the coordinate axes, say the x-axis. In the notation of Fig. 12, where θ is taken as the angle between the normal dn and the x-axis, the x-component of grad p is

$$(\cos \theta)(\partial p/\partial n) = (\cos \theta)(\mathrm{d}p)/(\mathrm{d}n) = (\mathrm{d}p)/(\mathrm{d}x), \tag{30}$$

since dn = $(\cos \theta)$ dx where dx is the distance in the x-direction between the surfaces of pressure p and $p+\mathrm{d}p$. But the discussion of eqn (23) identifies the right-hand side of eqn (30) as

$$\partial p/\partial x. \tag{31}$$

Similarly, for each coordinate the component of grad p in the direction of the associated coordinate axis is the partial derivative of p with respect to that coordinate.[†] The vector grad p, as specified through its components along the three coordinate axes, is therefore

$$\text{grad } p = \nabla p = (\partial p/\partial x, \partial p/\partial y, \partial p/\partial z). \tag{32}$$

Here we may note that eqn (29), giving the pressure force per unit volume as *minus* expression (32), can be independently derived by consideration of the particle illustrated in Fig. 9. The component of pressure force in the x-direction, for example, comes from the two faces perpendicular to that direction, and consists of minus the pressure difference (expression (23)) multiplied by the area dy dz of both those faces. On division by the particle's volume dx dy dz we obtain $-\partial p/\partial x$. This is quite a quick way of obtaining the pressure force per unit volume, but it yields less information about important matters such as magnitude and direction. In the rest of this book we constantly use eqn (29) to find the pressure force per unit volume, with grad p specified sometimes by expression (28) and sometimes by eqn (32).

[†] More generally, the resultant of grad p in *any* direction is equal to the rate of change of p in that direction.

Actually, the gravitational force per unit volume on a particle of fluid (ρg directed vertically downwards) can also be written as minus a gradient, namely as

$$- \operatorname{grad} (\rho g H), \tag{33}$$

where H represents height above some reference level (so that grad H is a vector of unit magnitude directed vertically upwards).[†] Taken with (29), this gives the total force per unit volume on a particle in the Euler model as

$$- \operatorname{grad} (p + \rho g H). \tag{34}$$

In the hydrostatic case, of course, eqn (3) holds and the total force in expression (34) is zero. For moving fluids, however, the total force per unit volume (expression (34)) is equal to the mass per unit volume ρ times the acceleration of the particle.

2.3 The velocity field and streamlines

Evidently, it is now of crucial importance to find a convenient expression for the acceleration of a fluid particle. Careful consideration soon makes it clear, however, that it is not at all convenient to follow the tortuous course of each individual particle of fluid through the pressure field, with its coordinates specified as functions $x(t)$, $y(t)$ and $z(t)$ of the time t so as to yield acceleration components $\ddot{x}(t)$, $\ddot{y}(t)$ and $\ddot{z}(t)$. This approach has been tried from time to time, by Euler and by a few of his successors (notably Lagrange, after whom the approach is usually named), but almost all of the successful developments in fluid mechanics have come from the approach preferred by Euler which is based on the concept of *the velocity field*.

The motion of a fluid, on this system, is specified entirely in terms of the distribution of the velocity in space and time. This means that the components (u, v, w) of velocity in the direction of the coordinate axes are viewed as functions

$$(u(x, y, z, t), \; v(x, y, z, t), \; w(x, y, z, t)) \tag{35}$$

of the coordinates x, y, and z, and of the time t.

This approach maintains absolutely no record of the history of individual particles of fluid! Rather, the dependence on the time t indicated in expression (35) is concerned exclusively with the history of conditions at a particular point

---

[†] As in Section 1.3, we can here interpret $\rho g H$ as potential energy per unit volume, and thence understand the form of expression (33) from the fact that if the particle moves a short distance in any direction the component of force in that direction multiplied by the short distance moved must be the work done (or minus the change of potential energy). This agrees with expression (33) since by the footnote on p. 19 the component of grad $(\rho g H)$ in any direction is the rate of change of $(\rho g H)$ with distance travelled in that direction.

in space. In the meantime, for any fixed value of t, the expression (35) specifies a *vector field*: the spatial distribution of the velocity vector (u, v, w) found at that time t.

One particular type of fluid flow turns out to be of great importance for the representation to good approximation of a very wide class of real motions of fluids. This is the type of flow in which the velocity components in expression (35) show no actual variation with time whatsoever. The expression *steady flow* is defined as meaning a flow with this property, which implies of course that the velocity field (*spatial* distribution of velocity) takes a form

$$(u(x, y, z), \ v(x, y, z), \ w(x, y, z)) \tag{36}$$

which is the same at all times.

The idea of steady flow has been encountered already in Section 1.2. In the steady flow through a pipe studied there the velocity varied with position (becoming larger in the narrower section of the pipe) but its value was unchanging at any fixed point, as implied by expression (36).

Similarly, in many other important areas of application of fluid mechanics (see the Preface), many flows are found to be effectively steady. Here, the qualification 'effectively' means that the velocity at a fixed point remains constant to a rather close approximation or takes constant values on which are imposed some relatively small fluctuations which may, for certain purposes, prove not to be important.

A steady flow in this sense may be observed, for example, by leaning over a bridge and observing the pattern where the river flow narrows to make its way between the piles of the bridge. If an individual particle of fluid is watched it is seen to be accelerating, but if a firm effort is made to concentrate the gaze on a fixed point in space the magnitude and direction of the fluid velocity at that point are seen to remain practically constant.

An observer leaning over the side of a ship (near the bow) as it moves through calm water sees quite a different (but also very interesting) type of steady flow. An essential feature of this is that, from the standpoint of the observer (that is, in a frame of reference fixed in the ship) the ship is at rest and the water is flowing steadily past it.[†]

A reader who attempts this particular observation may find a need to concentrate the gaze rather accurately on a fixed point (use of a sighting marker firmly attached to the ship can be recommended, and the eye needs to be held still). The bow makes waves, in fact, and the observer is easily seduced into watching individual particles of fluid as they bump up and down over the waves. Disciplined observation shows, however, that the entire wave pattern moves unchangingly forward with the ship: in other words, the pattern is steady

---

[†] Similarly, from the standpoint of an observer on an aircraft in flight, the wings of the aircraft are at rest in a steady airstream. Important features of such steady airflow patterns over wings are studied in Chapter 11.

in a frame of reference fixed in the ship and, at a point fixed relative to the observer (a point which may continue always to be on the forward face, or always on the backward face, of one of the waves), the magnitude and direction of the fluid velocity remain practically constant. Many further examples of steady flow appear later in this book; which, however, is concerned equally with developing the theories of both steady and unsteady flow.

We note in the meantime that one difficulty which generally characterises the velocity-field approach (namely the absence of information about the history of individual particles) disappears in the case of steady flow. If, in the velocity field of expression (36), we draw large numbers of very short arrows, each of length proportional to the local value of the velocity vector (Fig. 13), we can readily see how these arrows may be joined up to form an assemblage of curves called the *streamlines*.

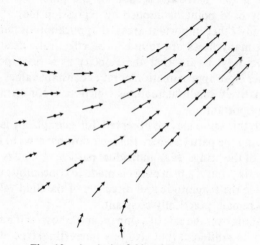

Fig. 13. A velocity field and its streamlines.

The relationship between a velocity field and its streamlines is identical with the relationship between an electric or magnetic field and its lines of force. Furthermore, just as the pattern taken up by iron filings in a magnetic field can be used to indicate the lines of force, so a similar experimental indication can be used for streamlines. If bright particles (e.g. of aluminium powder) are suspended in a flowing liquid and a flash photograph of its flow is taken then, during the exposure time, each particle moves a distance proportional to the local fluid velocity. The eye sees the resulting short streaks on the photograph as being joined up into curves which are projections of the streamlines on to the plane of the picture.

Mathematically, the streamlines associated with the velocity field of expression (36) represent the doubly infinite set of solutions of the differential equations

$$\frac{dx}{u} = \frac{dy}{v} = \frac{dz}{w},\tag{37}$$

where each expression in eqn (37) represents the very short time dt during which a particle of fluid makes the change of position (dx, dy, dz). Since the time does not appear explicitly in the velocity field of expression (36) it is, in principle, possible to treat eqn (37) as a pair of ordinary differential equations, e.g. for y and z in terms of x, the solutions of which (involving a pair of arbitrary constants) form a doubly infinite assemblage of streamlines.

Evidently, the streamlines of a steady flow are the paths along which fluid particles move. In fact, a particle on any one streamline remains always on that streamline because, as eqn (37) shows, its velocity vector (u, v, w) is everywhere directed along the curve. This view of the velocity field as defining an assemblage of curves, the streamlines, along which particles of fluid necessarily move, represents a most valuably simplifying feature of steady flows.

However, this book is concerned with the theory of both steady and unsteady flows. It is essential, therefore, to consider whether any features with properties partly analogous to those of streamlines exist for unsteady flow.

The answer is that *at any one instant* a pattern of streamlines can be defined exactly as in Fig. 13. It represents a set of curves such that, at that one instant, each particle of fluid is moving tangentially to the curve on which it lies. Experimentally, these would still be the curves indicated by a single flash photograph of a suspension of bright particles but such a photograph is in this case just a 'snapshot' picture of a streamline pattern which is itself continuously changing. Mathematically, the differential eqns (37) still define the streamlines, but u, v, and w are specified as in expression (35) and the time t is held constant while the equations are being solved.

In unsteady flow, the 'pathlines' (paths of particles of fluid) are completely different in shape from the shapes of the streamlines at any one instant. As a particle moves along a pathline it is at each instant moving tangentially to its local streamline, but the pattern of those streamlines is changing in time.

Experimentally, a long-duration exposure of a single bright particle suspended in the fluid would indicate its pathline. Mathematically, it would be necessary to solve a system of three ordinary differential equations

$$dx/dt = u, \ dy/dt = v, \ dz/dt = w,\tag{38}$$

with u, v, and w specified by expression (35), to determine the pathline. However, any attempts to carry out this complicated process (leading to a *triply infinite* set of solutions) have been largely avoided by workers in theoretical fluid mechanics.[†] Streamlines, on the other hand, have quite often (and quite usefully) been determined for unsteady flows.

2.4 The solenoidal property

One of the velocity field's very important properties can be explained most clearly if we extend slightly the concept of streamlines described in Section 2.3 and introduce the closely related concept of a streamtube. This is defined as a tube made up of streamlines. Given any loop (that is any closed curve) in space, the streamlines passing through points of the loop compose a surface which is called a streamtube (Fig. 14).

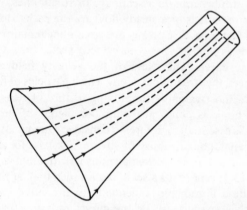

Fig. 14. A streamtube is a surface comprising all the streamlines passing through points of a particular loop.

In steady flow, of course, a streamtube has unchanging shape, and it confines the flow exactly like the pipes studied in Section 1.2. This is because the motion of each particle of fluid on its boundary is directed along the streamline on which it is situated and this lies in the bounding surface of the streamtube. That particle's motion is, in short, tangential to the streamtube surface, so there is no flow through that surface and flow inside the tube remains confined within it.

The analogy with the pipe flows of Section 1.2 becomes still stronger if we concentrate on an extremely *thin* streamtube (Fig. 15). This brings us very close indeed to the conditions assumed for the one-degree-of-freedom model of

[†] One more type of curve which can be made visible in a particular experimental arrangement is that sometimes called a streakline (or filament line). It is the locus of those particles of fluid that have passed through one individual point in space. A streakline is made visible by emitting dye from a hypodermic needle with its tip situated at that point. Streaklines coincide with streamlines in steady flow, but are quite different in unsteady flow from *both* streamlines and pathlines.

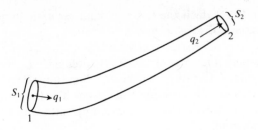

Fig. 15. Flow of fluid along an extremely thin streamtube.

Section 1.2. In fact, the flow velocity takes an almost uniform value over the cross-section of this streamtube (because the assumed continuity of its properties allows no significant variation over a very short distance).[†]

We write

$$q = (u^2 + v^2 + w^2)^{\frac{1}{2}} \qquad (39)$$

for the magnitude of the fluid velocity. This is the quantity which can be taken as effectively constant across a sufficiently thin streamtube and which represents the speed of flow through the tube. Now, if we consider two cross-sections of the streamtube (no. 1, with area S_1 and with $q = q_1$, and no. 2, with area S_2 and with $q = q_2$) then the arguments leading to eqn (5) give

$$S_1 q_1 = S_2 q_2. \qquad (40)$$

In other words, a very thin streamtube has a cross-sectional area S whose product with the flow speed q takes a uniform value

$$Sq = \text{constant} \qquad (41)$$

all along it. This constant value represents the rate of volume flow through the streamtube.

Although the truth of eqn (41) can be seen most directly in the case of steady flow, by analogy with flow through a pipe, we now show it to be correct also for quite general unsteady flows. At each instant, of course, a streamline pattern exists for an unsteady flow and this allows us to define a streamtube as in Fig. 14. Instantaneously, the motion of each particle of fluid on the surface of the tube is directed tangentially to the streamline on which it is situated and therefore tangentially to the tube composed of those streamlines. At the instant in question, then, there is no component of flow through the tube surface.

Applying this to the case of a very thin streamtube (Fig. 15), we see that the

[†] Also, the difficulties due to friction between the moving fluid and a stationary solid boundary which were mentioned in Section 1.3 are absent in the flow of Fig. 15, where the bounding surface is *not* a stationary solid but the moving fluid. These considerations allow us to anticipate a conclusion which is more fully substantiated in Section 3.5: that in every steady flow the total head will be found to take a constant value along any thin streamtube.

instantaneous rate at which the lump of fluid between stations 1 and 2 is changing volume comprises a positive contribution $S_2 q_2$ from station 2 where the lump is taking in additional volume and a negative contribution $- S_1 q_1$ from station 1 where it is vacating previously occupied volume. For a fluid of constant density ρ the net rate of change of volume must be zero, yielding eqn (40) as before. In short, there is a uniform volume flow (see eqn (41)) along the instantaneously existing streamtube because its surface is allowing no flow across it at that instant.

The important result (eqn (41)), that the magnitude of the velocity field q varies along any thin streamtube *in inverse proportion to its cross-sectional area* S (whether in steady, or in unsteady, flow) is called the solenoidal property of the velocity field (from the Greek *solen*, a tube). This is a property which it shares with other well known vector fields, such as the 'magnetic flux vector' of electromagnetic theory, or the 'electric flux vector' in a region without electrical charges. One reason for the property's importance is that it allows us to use the streamline pattern (or the pattern of lines of force as the case may be) to infer not only the *direction* of the vector field, but also its magnitude. That magnitude q is large where the streamlines bunch closely together and is small where they spread apart. More precisely, it varies in such a way that Sq takes a uniform value all along a thin streamtube.

Many readers will be aware from general vector field theory that the solenoidal property can also be expressed in the form of a partial differential equation,

$$\frac{\partial u}{\partial x} + \frac{\partial u}{\partial y} + \frac{\partial w}{\partial z} = 0. \tag{42}$$

We can, indeed, readily show that this equation, too, expresses mathematically the fact that there is zero rate of change of volume for a particle of fluid (assumed to be of constant density ρ) under those conditions of *continuity* explained in Section 2.1, conditions which have led to eqn (42) being universally known as the equation of continuity for an incompressible fluid.

Actually, we can go further and find a specific physical interpretation of the left-hand side of eqn (42) which, besides explaining that equation, becomes of importance when we begin (in Section 3.1) to extend the Euler model to a compressible fluid. In fact, for a general fluid (which may be compressible or incompressible) we consider the rate of change of volume of a particle per unit volume which is produced by a given velocity field (expression (35)).

For the particle in Fig. 9, for example, this consists of three terms: one associated with each of the three components (expression (35)) of the velocity. The x-component u of velocity acts only at the faces of area $dy\,dz$ which are perpendicular to the x-direction to generate changes in the volume occupied by the particle. The excess of new volume per unit time occupied at $x + dx$ (the right-hand face) over old volume vacated at x (the left-hand face) depends on the excess velocity

$$(\partial u / \partial x)\, \mathrm{d}x \qquad\qquad (43)$$

found at any point on the right-hand face in comparison with the corresponding point on the left-hand face. The associated rate of change of volume is expression (43) multiplied by the area $\mathrm{d}y\,\mathrm{d}z$ which, for a particle of volume $\mathrm{d}x\,\mathrm{d}y\,\mathrm{d}z$, gives a rate of change of volume per unit volume of $\partial u / \partial x$. Similar arguments for the contributions from the y- and z-components determine the total rate of change of volume for the particle per unit volume to be the left-hand side of eqn (42).

Besides establishing eqn (42) as the correct equation of continuity for an incompressible fluid, the above argument introduces a quantity of great importance for any vector field. To represent a vector field we adopt the usual bold type-face, thus:

$$\mathbf{u} = (u, v, w), \text{ with magnitude } q. \qquad\qquad (44)$$

For a completely general velocity field $\mathbf{u}$ we define the *divergence* of $\mathbf{u}$ (written as div $\mathbf{u}$ or sometimes as $\nabla \cdot \mathbf{u}$) as the resulting rate of change of volume per unit volume

$$\operatorname{div} \mathbf{u} = \nabla \cdot \mathbf{u} = \frac{\partial u}{\partial x} + \frac{\partial v}{\partial y} + \frac{\partial w}{\partial z}. \qquad\qquad (45)$$

This definition allows us to write down the rate of change of volume for a large lump L of fluid as

$$\int_{L} \operatorname{div} \mathbf{u}\, \mathrm{d}V, \qquad\qquad (46)$$

an expression which envisages the lump to be composed of small particles of volume $\mathrm{d}V$ (which is equal, for a particle shaped as in Fig. 9, to $\mathrm{d}x\,\mathrm{d}y\,\mathrm{d}z$) each of which therefore has a rate of change of volume div $\mathbf{u}\,\mathrm{d}V$. The integral of this represents the rate of change of volume of the whole lump. An alternative expression for this considers the bounding surface ∂L of the lump (Fig. 16) and the rate at which the component of velocity normal to that surface (which we write $\mathbf{u} \cdot \mathbf{n}$ in terms of $\mathbf{n}$ which is a unit vector along the outward normal to the surface) changes the volume of the lump. Where $\mathbf{u} \cdot \mathbf{n}$ is positive, it represents the rate at which the lump moves to occupy new volume per unit area of the surface. Where it is negative it represents the rate of change associated with vacating previously occupied volume. Therefore, the integral

$$\int_{\partial L} \mathbf{u} \cdot \mathbf{n}\, \mathrm{d}S, \qquad\qquad (47)$$

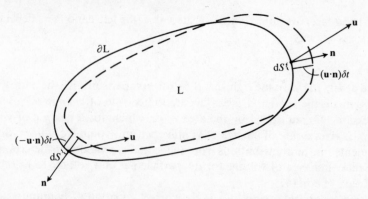

Fig. 16. In a velocity field **u** the lump of fluid L, with boundary ∂L given by the plain line, moves after time δt to occupy the region bounded by the broken line. At the right-hand boundary element of area dS, where **u · n** is positive (**n** being a unit vector along the outward normal to L), the movement increases the lump's volume by $[(\mathbf{u \cdot n})\, dt]\, dS$. At the left-hand boundary element of area dS, where **u · n** is negative, the volume change takes the same form but of course represents a decrease.

where dS is a small area element of the bounding surface ∂L, represents the total rate of change of volume and must be equal to expression (46).

The identity of expressions (46) and (47), which was established above for velocity fields, is the well known general property of vector fields which is proved analytically in mathematics texts and which is known as the Divergence Theorem:

$$\int_{L} \text{div } \mathbf{u}\, dV = \int_{\partial L} \mathbf{u \cdot n}\, dS. \tag{48}$$

It equates the volume integral of the divergence over a region L to a surface integral of the vector field's component along the outward normal **n** over the boundary surface ∂L.

As a first illustration of the use of this theorem (which is applied frequently in subsequent chapters) we consider the lump L between the cross-sections S_1 and S_2 in Fig. 15. Assuming the equation of continuity for an incompressible fluid, eqn (42), the left-hand side of eqn (48) is zero. The integral on the right-hand side comprises a positive contribution $S_2 q_2$ from cross-section S_2 (where the outward normal component **u · n** is $+q_2$) and a negative contribution $-S_1 q_1$ from cross-section S_1 (because the velocity there is directed along the inward normal). There is no contribution from the tube surface itself because the fluid flow is purely tangential to that surface. This establishes the solenoidal property (eqn (40)) as directly equivalent to the equation of continuity (eqn (42)).

3
Equations of motion

The groundwork laid in Chapter 2, on the assumptions underlying the Euler model and on the basic properties of pressure fields and velocity fields, makes possible the formulation of equations of motion that fully determine, on the Euler model, the motion of a fluid. The present chapter obtains these equations of motion both for the Euler model as defined in Chapter 2 and for a model taking variability of density into account; even though no use of the latter model is made in the rest of this book.

In carrying out this programme two main difficulties arise: one mathematical and one physical. When the principles of mechanics are applied to a particle of fluid, a mathematical expression for its acceleration is needed. This demands a convenient formulation of the rate of change following a particle of fluid, as we first investigate. The consequent expression of that momentum principle, foreshadowed in Section 2.1, takes the form of a further partial differential equation to be satisfied along with the equation of continuity (eqn (42)). But in order to determine the pressure and velocity fields a statement not only of partial differential equations but also of corresponding boundary conditions is required; and the formulation of the boundary conditions appropriate to the Euler model raises important physical difficulties which need to be discussed and elucidated in some detail.

3.1 Rate of change following a particle

We begin by determining how the equation of continuity (eqn (42)) is changed in a model that takes variability of density into account. This material is included partly for its own intrinsic interest, but mainly because it forces us to obtain a mathematical expression for the rate at which a particle's density ρ is changing. This discussion of the rate of change of the scalar field ρ following a particle is useful as a preliminary to the analysis of that particle's acceleration (rate of change of the vector field $\mathbf{u}$ following the particle).

The equation of continuity (eqn (42)) expresses the situation when a small particle of fluid has an unchanging volume. If, however, the fluid has variable density ρ then the particle's volume V must change in such a way that the product ρV (the mass of the particle) remains constant. This means that the rate of change of ρV for the particle is zero:

$$V \, d\rho/dt + \rho \, dV/dt = 0. \tag{49}$$

Equation (45) specifies the particle's rate of change of volume per unit volume as the divergence of the velocity field

$$V^{-1}dV/dt = \operatorname{div} \mathbf{u}. \tag{50}$$

With eqn (49), this gives

$$d\rho/dt = -\rho \operatorname{div} \mathbf{u}. \tag{51}$$

Although the equation of continuity for a compressible fluid may be written in the relatively simple form of eqn (51), careful attention must be paid to the interpretation of the left-hand side. This is the rate of change of density for the particle of fluid as it moves along its instantaneous streamline.

If we express the density as a scalar field

$$\rho(x, y, z, t) \tag{52}$$

we might at first expect that the partial differential coefficient $\partial\rho/\partial t$ must be the required rate of change of density for the particle of fluid. This would be a mistake, however, since the partial differential coefficient $\partial\rho/\partial t$ is, of course, defined for the function (52) as the rate of change of ρ with respect to t when the other variables x, y and z are kept constant. Thus,

$$\partial\rho/\partial t \tag{53}$$

represents *local rate of change*; it is the rate of change of ρ at a fixed point (x, y, z) in space, but it is not the rate of change for a particle which, in general, is moving away from that point along its instantaneous streamline.

If dt represents a very short time interval (in the sense of the footnote on p. 14), then the particle situated at the point (x, y, z) at time t must be at the point

$$(x + u\,dt, \ y + v\,dt, \ z + w\,dt) \tag{54}$$

at time $t + dt$ (since u, v and w are rates of change of its x-, y- and z-coordinates with time). The corresponding change $d\rho$ in the density field (52) is therefore

$$\rho(x + u\,dt, \ y + v\,dt, \ z + w\,dt, \ t + dt) - \rho(x, y, z, t). \tag{55}$$

The theory of partial differential coefficients tells us that for small enough dt the small change (55) can be represented as a linear combination of four contributions

$$(u\,dt)\partial\rho/\partial x + (v\,dt)\partial\rho/\partial y + (w\,dt)\partial\rho/\partial z + (dt)\partial\rho/\partial t \tag{56}$$

due to the changes in each of the four variables x, y, z and t, respectively. The ratio of this change $d\rho$ to the small time interval dt can therefore be written

$$d\rho/dt = u\partial\rho/\partial x + v\partial\rho/\partial y + w\partial\rho/\partial z + \partial\rho/\partial t. \tag{57}$$

Physically, we can understand the first three terms on the right-hand side of eqn (57) if we write them compactly as the scalar product

$$\mathbf{u} \cdot \nabla\rho \tag{58}$$

of the velocity vector in eqn (44) with the gradient vector grad ρ or $\nabla\rho$ defined in eqn (32). Now, the scalar product $\mathbf{u} \cdot \mathbf{a}$ of $\mathbf{u}$ with any vector $\mathbf{a}$ is defined so as to be equal to the magnitude q of $\mathbf{u}$ times *the resultant of $\mathbf{a}$ in the direction of $\mathbf{u}$*; that is, its resultant along the instantaneous streamline. Thus, we can write (58) as

$$q \partial \rho / \partial s, \tag{59}$$

where $\partial \rho / \partial s$ is the rate of change of ρ with distance along the instantaneous streamline (namely, as explained in the footnote on p. 19, the resultant of grad ρ in that direction).

The expression (58) (or (59)) is called the *convective rate of change* of ρ. It is the rate of change due solely to the particle's changing position. In a small time dt, the particle moves a distance $ds = q\,dt$ along its instantaneous streamline and this contributes a change $(\partial \rho / \partial s) q\,dt$ to $d\rho$, or $q \partial \rho / \partial s$ to $d\rho/dt$.

However, the *total* rate of change $d\rho/dt$ for the particle consists of the convective rate of change, expression (59) or (58), due to change of position, plus the *local* rate of change (expression (53)) due to any variability of the density field of expression (52) at a fixed position. Actually, the phrase 'total' rate of change of ρ is commonly used in theoretical fluid mechanics to signify this sum of the local rate of change $\partial \rho / \partial t$ and the convective rate of change $\mathbf{u} \cdot \nabla \rho$ which constitutes the rate of change following a particle.

On the other hand, any tendency to use such a general mathematical symbolism as $d\rho/dt$ for total rate of change has increasingly been abandoned in theoretical fluid mechanics. There is no general mathematical concept involved here (as there is for the divergence operator in eqn (45), which can be applied to electromagnetic and other vector fields as well as to velocity fields). Only in fluid mechanics, with $\mathbf{u}$ as the velocity field, does the total rate of change in eqn (57) have a physically important meaning (as rate of change following a particle). Therefore, a notation $D\rho/Dt$ special to fluid mechanics is commonly used for this quantity.

To sum up, for the density ρ, or for any scalar field, the total rate of change $D\rho/Dt$ is defined as the rate of change of ρ following a particle of fluid:

$$D\rho/Dt = \partial \rho / \partial t + \mathbf{u} \cdot \nabla \rho, \tag{60}$$

which is the sum of the local and convective rates of change of ρ. On the basis of this definition, the equation of continuity (eqn (51)) for a compressible fluid can be quite precisely written as

$$D\rho/Dt = -\rho \operatorname{div} \mathbf{u}. \tag{61}$$

3.2 Acceleration of a particle of fluid

In Section 3.1 we defined an operator

$$\frac{D}{Dt} = \frac{\partial}{\partial t} + \mathbf{u} \cdot \nabla \tag{62}$$

which, for any velocity field $\mathbf{u}$, can be applied to any scalar quantity (for example, to ρ in eqn (60)) to give the rate of change of that quantity following a particle of fluid. Now a particle's acceleration, of course, is its rate of change of velocity; but, at first sight, two difficulties stand in the way of applying the operator in eqn (62) to a particle's velocity in order to obtain its acceleration.

The first difficulty is that eqn (62) defines an operator to be applied to scalar quantities only (particularly since in Section 2.2 we defined the gradient, grad or ∇, as an operator applied to scalar quantities only); yet a particle's velocity is a vector. This difficulty disappears when we remember that the rate of change of each component u, v, or w of the particle's velocity is equal to the corresponding component of acceleration; for example, the z-component of acceleration is Dw/Dt, which is the effect of applying the operator in eqn (62) to the scalar quantity w. The particle's acceleration, then, should be a vector whose components are obtained by applying the operator in eqn (62) to the three components (u, v, and w) of velocity.

In theoretical fluid mechanics this vector is usually denoted by $D\mathbf{u}/Dt$ even though eqn (62) strictly defines an operator that can be applied only to scalars. That definition is extended, in fact, by further defining

$$D\mathbf{u}/Dt = (Du/Dt, Dv/Dt, Dw/Dt). \tag{63}$$

The immediate appearance of eqn (63), however, seems to raise another difficulty: in every case, the definition of the *operator* in eqn (62) involves the velocity field $\mathbf{u}$, and this might be thought to cast doubt on whether that operator can be applied to the velocity field itself! Here, there is no real difficulty to be overcome; the above proof that the three components of the acceleration vector are as specified in eqn (63) is rigorously correct. However, the form of eqn (63) demonstrates an important difference between theories of velocity fields and other classical (e.g. electromagnetic) field theories: fluid mechanics, in fact, is governed by *nonlinear* equations because the acceleration field depends nonlinearly on the velocity field.

The character of the nonlinearity is as follows: the acceleration vector can be written

$$D\mathbf{u}/Dt = \partial\mathbf{u}/\partial t + \mathbf{u} \cdot \nabla\mathbf{u}, \tag{64}$$

where (once again) the vector $\mathbf{u} \cdot \nabla\mathbf{u}$ (a form at first sight undefined since the gradient operator ∇ is to be applied only to scalar fields) is defined as

$$\mathbf{u} \cdot \nabla\mathbf{u} = (\mathbf{u} \cdot \nabla u, \mathbf{u} \cdot \nabla v, \mathbf{u} \cdot \nabla w). \tag{65}$$

From (say) the first two components of eqn (65) written out in full as

$$u\partial u/\partial x + v\partial u/\partial y + w\partial u/\partial z, \; u\partial v/\partial x + v\partial v/\partial y + w\partial v/\partial z, \qquad \cdot \tag{66}$$

we observe how the acceleration field depends *quadratically* on the velocity field.

The components in expression (66) allow us, incidentally, to disprove one tempting conjecture. Since for a scalar field p we can by eqn (32) equate $p\nabla p$ to $\nabla(\frac{1}{2}p^2)$, we might suppose that $\mathbf{u} \cdot \nabla\mathbf{u}$ can be equated to $\nabla(\frac{1}{2}\mathbf{u} \cdot \mathbf{u})$. This is seen to be incorrect, however, when we write down the first two components of $\nabla(\frac{1}{2}\mathbf{u} \cdot \mathbf{u})$ as

$$u\partial u/\partial x + v\partial v/\partial x + w\partial w/\partial x, \; u\partial u/\partial y + v\partial v/\partial y + w\partial w/\partial y. \tag{67}$$

This example emphasizes that we must proceed with due care and scepticism when applying nonlinear operators to vector fields.

3.3 Momentum principle for a particle

The momentum principle applied to any particle takes the form of Newton's second law of motion: the particle's mass times its acceleration equals the total force acting on the particle. That total force on a particle, per unit volume, is given in the Euler model as expression (34), while the particle's mass, also per unit volume, is the density ρ. Equation (64) defines its acceleration. The momentum principle therefore takes the form

$$\rho\,D\mathbf{u}/Dt = -\operatorname{grad}(p + \rho gH). \tag{68}$$

The complete set of partial differential equations defining the Euler model can now be written down, with the right-hand side of eqn (64) substituted for $D\mathbf{u}/Dt$ in eqn (68), and the equation of continuity (42) written in terms of the divergence in eqn (45), as

$$\nabla \cdot \mathbf{u} = 0, \; \rho(\partial\mathbf{u}/\partial t + \mathbf{u} \cdot \nabla\mathbf{u}) = -\nabla(p + \rho gH). \tag{69}$$

These equations constitute one scalar and one vector equation for one unknown scalar field p and one unknown vector field $\mathbf{u}$. Alternatively, in terms of the three components of the vector equation, they constitute four scalar equations for four scalar variables p, u, v, and w.

The form in which the pressure p occurs in eqns (69) suggests that a notation

$$p_e = p + \rho gH - p_o = p - (p_o - \rho gH) \tag{70}$$

may occasionally be useful. Here, the constant p_o signifies the pressure at the reference level $H = 0$ in purely hydrostatic conditions, so that $p_o - \rho gH$ represents the purely hydrostatic pressure distribution as given by eqn (3). Accordingly, p_e is an *excess* pressure (which of course may be positive or negative) due to motion; it represents any excess of the pressure field over its hydrostatic value. The momentum equation (68) takes the form

$$\rho\,D\mathbf{u}/Dt = -\operatorname{grad} p_e \tag{71}$$

in terms of the excess pressure p_e (also called 'reduced' pressure in some texts).

The right-hand side of eqn (71) represents the total force (per unit volume) on a particle.

We conclude this section by noting briefly the form

$$D\rho/Dt = -\rho\nabla\cdot\mathbf{u}, \rho D\mathbf{u}/Dt = -\nabla p - \rho\nabla(gH) \qquad (72)$$

taken by the equations of motion of a fluid of variable density ρ. The first of these is the equation of continuity for a compressible fluid, eqn (61); the second equates mass times acceleration per unit volume to force per unit volume (note that the gravitational force per unit volume, $-\rho\nabla(gH)$, can only be written as in eqn (69) when the density ρ is constant).

Equation (72) constitute four scalar equations in *five* unknowns, ρ, p, u, v, and w, so that they need to be supplemented with a fifth equation expressing the compressibility properties of the fluid. They are of widespread utility because they underlie theories of the propagation of sound waves, as well as being needed, as noted in Section 2.1, to study any flows of air at speeds of the order of 100 m/s or more. However, they are not further referred to in the remainder of the present book.

3.4 Boundary conditions

Because the eqns (69) governing the mechanics of a fluid in the Euler model are partial differential equations, we must consider with care what additional information is needed in order to determine fully a particular motion. For all mechanical systems, of course, we expect to have to specify *initial conditions*. In general, the eqns (69) are no exception: the equation giving $\partial\mathbf{u}/\partial t$ tells us about the rate of change of the velocity field of expression (35) at each instant and therefore can fully determine the motion only if an initial velocity field is known.

Initial conditions are needed in general, then, although in fluid mechanics there is one important exception to that rule. This is the case when we seek to determine a steady flow, specified by a velocity field of the form of expression (36) independent of t, which satisfies the eqns (69). Some special features of the equations in this case are noted later in the present book, beginning with Section 3.5. The essential point is that the $\partial\mathbf{u}/\partial t$ term in eqns (69) is missing—this, indeed, is why no specification of initial conditions is needed.

Partial differential coefficients with respect to x, y, and z occur in eqns (69) (as is clear from the eqn (45) for $\nabla\cdot\mathbf{u}$ and expressions (66) for components of $\mathbf{u}\cdot\nabla\mathbf{u}$) whether the flow is steady or unsteady. Just as an equation specifying the time rate of change $\partial\mathbf{u}/\partial t$ needs complementing by an initial condition if the motion is to be fully determined, so equations involving spatial differential coefficients need complementing by conditions to be satisfied at the fluid's spatial boundary. We now discuss carefully the *boundary conditions* appropriate to the Euler model.

All fluids possess solid boundaries. Examples of these for the fluid motions noted in the Preface are as follows: external surfaces of animals, or of man-made vehicles, moving through a fluid; internal surfaces of constraining pipes or tubes; the surface of the ground or of the sea bottom for atmospheric or oceanic motions; external surfaces either of blades in engines, or of structures designed to resist wind and wave. We need to devote special attention to the boundary conditions appropriate at such solid boundaries.

Liquids possess, in addition, *free* boundaries (unlike gases which necessarily fill any solid container in which they are placed). The boundary conditions at a free boundary are found, however, to be relatively easy to specify (see the end of this section) once the conditions at a solid boundary have been properly analysed.

There is one obvious condition that clearly must be satisfied at a solid boundary; in simple words, this is that there is no fluid flow *through* the solid surface. This is a restriction on the component of fluid velocity normal to the surface (the question of whether there is or is not a component tangential to a solid boundary must be carefully considered later).

As in Section 2.4 we use $\mathbf{n}$ for a unit vector along the outward normal to the fluid's boundary (that is, outward from the fluid and thus into the solid). If the solid boundary is at rest, the condition of zero normal flow through it can be written

$$\mathbf{u} \cdot \mathbf{n} = 0. \tag{73}$$

From the standpoint of Section 2.1 this is a further expression of *continuity*: it states that there is a zero rate of disappearance of fluid, or of creation of new fluid, at a solid boundary.

We are also interested in moving solid boundaries such as turbine blades. In this case, the absence of flow through the boundary depends on the *relative* velocity between fluid and solid having a zero normal component. If $\mathbf{u}_s$ is the velocity of the solid, this relative velocity is $\mathbf{u} - \mathbf{u}_s$ and the boundary condition is therefore

$$(\mathbf{u} - \mathbf{u}_s) \cdot \mathbf{n} = 0. \tag{74}$$

A certain alternative form of eqn (74) which is very often useful involves an interesting application of the operator D/Dt defined in eqn (62). It can be applied if the geometrical equation of the solid surface in Cartesian coordinates x, y, z is known at each time t, in (say) the form

$$F(x, y, z, t) = 0. \tag{75}$$

(Clearly, the separate dependence on t in eqn (75) would disappear only if the solid boundary were at rest.)

We consider now the rate of change DF/Dt following a particle of fluid on the solid boundary. The condition in eqn (74) means that the particle's motion relative to that of the solid surface is purely tangential to the boundary (it has zero normal component). Thus, the particle moves *along* the solid surface and the value of $F(x, y, z, t)$ following the particle continues to be zero (as specified by the eqn (75) of the surface). In short, there is zero rate of change of F for a particle at the surface:

$$0 = DF/Dt = \partial F/\partial t + \mathbf{u} \cdot \nabla F. \tag{76}$$

Although the conditions in eqns (74) and (76) look so different they are, in fact, mathematically equivalent. Equation (28) defines the gradient vector ∇F as a vector of magnitude $\partial F/\partial n$ in the direction F increasing. The scalar product $\mathbf{u} \cdot \nabla F$ can therefore be written as

$$(\partial F/\partial n)(\mathbf{u} \cdot \mathbf{n}); \tag{77}$$

that is, as the magnitude of ∇F times the component of $\mathbf{u}$ in the direction of ∇F. But if in time dt the surface $F = 0$ moves, normal to itself, a distance $(\mathbf{u_s} \cdot \mathbf{n}) dt$, the change of F at the surface is the sum of two terms

$$(\partial F/\partial t)dt + (\partial F/\partial n)(\mathbf{u_s} \cdot \mathbf{n}) dt \tag{78}$$

resulting from variation with time at a fixed position and from the normal displacement of position $(\mathbf{u_s} \cdot \mathbf{n}) dt$. This change must be zero (since $F = 0$ defines the surface), giving $\partial F/\partial t$ as

$$- (\partial F/\partial n)(\mathbf{u_s} \cdot \mathbf{n}); \tag{79}$$

so that eqn (76) requiring the sum of expressions (79) and (77) to be zero is equivalent to the boundary condition's alternative expression, eqn (74).

The crucial assumption in the Euler model, as noted in Section 2.1, is that the fluid is *frictionless*. This assumption strongly suggests that the condition in eqn (74) for zero normal fluid velocity relative to a solid surface could not appropriately be supplemented by any restriction whatsoever on the fluid's tangential motion along the solid surface. In fact, we shall find that the Euler eqn (69), involving only first-order partial differential coefficients with respect to x, y, and z, need only the one condition, eqn (74), at a solid boundary in order to determine fully a flow which, in general, involves substantial tangential motion at the solid surface.

Many readers new to the study of fluid motions may see no particular difficulty in the fact that the Euler model implies substantial tangential motions at solid surfaces. Casual observation seems, indeed, to indicate that fluid motions often possess such substantial tangential components at solid boundaries.

Rigorous experimental study has proved, however, that this is an illusion. For any solid boundary, the fluid very close indeed to it is greatly retarded and the limiting value of the tangential velocity as the solid boundary is approached is always zero.

In cases where casual observation indicates vigorous tangential motion at a solid boundary, there is really a very thin layer of retarded motion separating that vigorous flow from the boundary itself. This is called the *boundary layer*. In this very thin boundary layer the tangential component of velocity is diminished continuously from its 'vigorous' mainstream value to a limiting value of zero at the solid surface.

It is, of course, friction that causes the retardation of fluid in the boundary layer near to the solid surface. As noted in Section 2.1, a steep gradient of velocity (such as occurs in a boundary layer) generates a tangential force per unit area equal to the viscosity μ times that velocity gradient. Often, boundary layers are *very* thin for fluids with small viscosity (like air and water) because the tangential force can be sufficient to generate the necessary retardation to zero at the surface itself *only* if the gradient of velocity is very large. The reader should note in this context that any discontinuity of velocity at the surface itself (that is, any tangential motion of the fluid relative to the solid) would locally involve an *infinite* gradient of velocity and therefore an infinite retarding force that would instantaneously annul such a discontinuity.[†]

The relationship of the Euler model, which neglects all effects of viscosity, to real fluid flows with very thin boundary layers (layers whose thickness tends to zero with the viscosity) is, then, that a flow calculated on the Euler model is a close representation of the flow *outside* the boundary layer. The large-scale motions of the fluid (which, for many purposes, are the most important) are thus given by the Euler model, and involve substantial tangential motion at the boundary. That model does not, however, identify the extremely small scale of the very thin boundary layer separating those substantial tangential motions from the real solid surface. In the language of physics it is like an insufficiently high-powered lens which fails to 'resolve' the scale of that very thin boundary layer.

For fluids of small viscosity, including air and water, many flows do involve only very thin boundary layers, so that all of their large scale features can be predicted well by the Euler model. This is particularly important because (as we shall see) flows that are well predicted by the Euler model possess features advantageous in many engineering applications.

Unfortunately, however, there are also very many flows for which it is impossible for the associated boundary layers to remain thin, however small the fluid's viscosity may be. The problem, as we shall see (in Section 5.6), is that for certain types of motion the mechanics of the boundary layer itself causes the

---

[†] It may be of interest to observe that, mathematically, the effect of taking viscosity into account is found to be the addition, on the right-hand side of eqn (68), of a new force term involving *second-order* partial differential coefficients (which, actually, we can write $\mu \nabla^2 \mathbf{u}$). This raising of the 'order' of the equation means that the boundary condition in eqn (74) needs to be significantly supplemented if the consequent motion is to be fully determined; indeed, in order to achieve this the stronger condition $\mathbf{u} - \mathbf{u}_s = 0$ has to be used.

whole flow to *separate* from the solid surface (in the manner illustrated already in Fig. 6), and that this separation yields a flow very different from that of the Euler model for any value of the viscosity.

Nevertheless, readers will find a grasp of fluid mechanics based on the Euler model (with eqn (74) as the boundary condition) useful provided that they also acquire insight into the conditions under which the flow predicted by the Euler model makes probable the avoidance of boundary layer separation. The present book is devoted to the communication of such insight, alongside ability to use the Euler model and appreciation of the advantages of certain types of motion which it predicts with good accuracy.

This section's discussion of boundary conditions has dealt at length with the difficult matters associated with conditions at a solid boundary. We conclude it by a much briefer reference to the conditions at the *free boundary* of a liquid in contact with the atmosphere (as in oceans, rivers or lakes). The shape of such a free boundary is, of course, one of the unknown quantities which a model seeks to determine. Thus, although in terms of an eqn (75) for that shape the boundary condition in eqn (76) must still be satisfied, nevertheless as F is now unknown we need an additional condition to provide, effectively, a boundary condition sufficing to determine u, v, w, and p.

Briefly, this extra condition at a free boundary is the condition (used once or twice already in Chapter 1) that the pressure p in the liquid is equal to the atmospheric pressure p_a. This free surface condition

$$p = p_a \tag{80}$$

is found to be adequate for the discussion of all motions with a free surface except those involving ripples of very small wavelength (less than 0.1 m), for which a correction is needed to allow for the action of surface tension in the balance of force per unit area acting on an element at the surface.

3.5 Some properties of steady flows

For steady flows, the velocity field takes a form like expression (36) which is independent of t, so that the eqns (69) governing the Euler model are simplified (through the disappearance of the $\partial \mathbf{u}/\partial t$ term) to

$$\nabla \cdot \mathbf{u} = 0, \quad \rho \mathbf{u} \cdot \nabla \mathbf{u} = -\nabla(p + \rho g H). \tag{81}$$

Another simplifying property of steady flows was noted in Section 2.3: the streamlines form a fixed set of curves along which particles of fluid move. We show now how to deduce from *either* of these simplifications a most important result, foreshadowed in the footnote on p. 25.

Operating first from eqns (81), we consider the scalar product of the velocity vector $\mathbf{u}$ given by eqn (44) with the acceleration vector $\mathbf{u} \cdot \nabla \mathbf{u}$ given by eqn (65).

That scalar product is the sum

$$u(\mathbf{u}\cdot\nabla u)+v(\mathbf{u}\cdot\nabla v)+w(\mathbf{u}\cdot\nabla w)=\mathbf{u}\cdot(u\nabla u+v\nabla v+w\nabla w)$$

$$=\mathbf{u}\cdot(\nabla\tfrac{1}{2}u^2+\nabla\tfrac{1}{2}v^2+\nabla\tfrac{1}{2}w^2), \tag{82}$$

since, as noted in Section 3.2, the definition (eqn (32)) makes $p\nabla p=\nabla\tfrac{1}{2}p^2$ for any scalar p. Curiously, then, we have shown that

$$\mathbf{u}\cdot(\mathbf{u}\cdot\nabla\mathbf{u})=\mathbf{u}\cdot\nabla(\tfrac{1}{2}q^2); \tag{83}$$

which, in words, says that the vectors for which two components of each were written out in (66) and (67) (in order to prove them to be different vectors) have been shown to have identical scalar products with the fluid velocity $\mathbf{u}$.

The scalar product of $\mathbf{u}$ with the second of the eqns (81) can be written, using eqn (83), as

$$\mathbf{u}\cdot\nabla(p+\rho gH+\tfrac{1}{2}\rho q^2)=0. \tag{84}$$

The quantity in brackets (pressure plus the sum of the potential and kinetic energies per unit volume of fluid) is the *total head* as defined in Section 1.3. Equation (84) states the famous result of Daniel Bernoulli (1700–82) that, in steady flow on the Euler model, the rate of change of total head for a particle of fluid as it moves along a streamline is zero. As in Section 1.3, this reflects the absence of any frictional dissipation of energy in the Euler model.

More directly, we can deduce it from a result in classical mechanics for the acceleration of any particle moving along a curve fixed in space (here, the streamline). That acceleration has two components: a component *along* the curve of $q\partial q/\partial s$ (rate of change of $\tfrac{1}{2}q^2$ in that direction); and, if the curve has radius of curvature R, a component q^2/R directed towards its centre of curvature. The first part of this statement makes the component of eqn (81) along the curve

$$\partial(p+\rho gH+\tfrac{1}{2}\rho q^2)/\partial s=0, \tag{85}$$

thus confirming the Bernoulli result that total head takes a constant value along any streamline.

Another way of writing this uses the definition (70) of the 'excess pressure' p_e to deduce that

$$p_e+\tfrac{1}{2}\rho q^2=\text{constant along a streamline.} \tag{86}$$

Similarly, from the form of eqn (71) for the momentum equation using p_e, together with the second part of the result from mechanics quoted above, we obtain the sometimes useful information that in steady flow the resultant of grad p_e normal to any streamline has the value

$$\rho q^2/R \text{ directed away from the centre of curvature.} \tag{87}$$

One limiting case of this should be noted: for a straight streamline R is infinite, so that the pressure gradient normal to it is zero in steady flow.

Three very simple inferences using the above principles may be noted. The *pitot-static* tube, illustrated schematically in Fig. 17, is designed to infer the velocity U of the steady stream from the measured difference between the pressure at a small orifice A facing the stream and that at an orifice B in the side of the tube. Excess pressures due to the tube disturbing the uniform motion with speed $q = U$ may be presumed negligible far upstream so that the quantity in eqn (86) takes the value $\frac{1}{2}\rho U^2$ on each and every streamline. The excess pressure p_e at A, where the oncoming flow is brought to rest ($q = 0$), is therefore $\frac{1}{2}\rho U^2$. At B, on the other hand, we may assume that the streamlines have become straight again so that there is no gradient of excess pressure normal to them, and the value of p_e measured at B is zero. The difference between the pressures at A and B therefore gives to good approximation the value of $\frac{1}{2}\rho U^2$.

The second of many examples where the effect of bringing to rest a steady stream of velocity U is to create a local excess pressure $\frac{1}{2}\rho U^2$ is that of the impact of a jet on a wall (Fig. 18), which has been mentioned already in Section 1.2. The constant in eqn (86) is once more $\frac{1}{2}\rho U^2$ on *each* streamline,

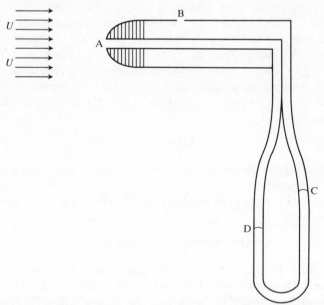

Fig. 17. Schematic illustration of the pitot-static tube. The value of the wind-speed U is derived from the difference, $\frac{1}{2}\rho U^2$, between the pressures at A and at B (a simple manometer, for example, would derive this from the difference in height of the liquid levels C and D).

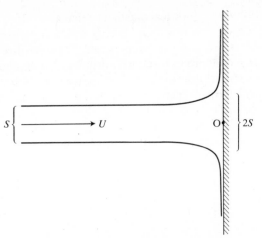

Fig. 18.   Impact of a jet on a wall. The force $\rho\,SU^2$ acting on the wall can be written as a product of the *maximum* excess pressure on the wall, $\frac{1}{2}\rho U^2$, and an area $2S$ equal to twice the jet cross-section.

because the excess pressure p_e is zero upstream where $q = U$. At the wall, then, the maximum excess pressure is $\frac{1}{2}\rho U^2$, which is achieved at the point O of the wall where the tangential velocity is zero. Knowing that the total force on the wall (from Section 1.2) is ρSU^2, we deduce that the excess pressures contributing to it (all $\leqslant \frac{1}{2}\rho U^2$) are significant over an area substantially greater than $2S$.

Finally, we note another case where combined knowledge of the solenoidality and Bernoulli properties of steady flow gives valuable qualitative, but insufficient quantitative, information. Such knowledge allows us to guess why a shape, as in Fig. 19, placed in the steady horizontal stream shown, sustains a 'lift' (vertically upward force). Once again, the constant in eqn (86) is seen, on the Euler model, to take a value $\frac{1}{2}\rho U^2$ on each streamline.[†] Immediately above

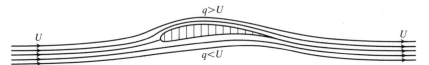

Fig. 19.   Schematic illustration of streamlines for steady flow around a 'cambered aerofoil' shape. This suggests qualitatively why such a flow may generate a vertically upward force or 'lift' on the aerofoil (to be studied quantitatively in Chapter 10).

[†] In the real flow, the total head takes a value less than this only on those streamlines that have passed through the thin 'boundary layer' (Section 3.4) where viscous action is causing dissipation of energy. Behind the body, these streamlines make up a thin 'wake' where the total head takes reduced values (associated with flow speeds $q < U$).

the body, streamlines are bunched together so that $q > U$ (Section 2.4) and the excess pressure p_e is negative. Immediately below it, streamtube areas are expanded so that $q < U$ and the excess pressure p_e is positive. (These results, incidentally, are seen to accord well with expression (87), since the centres of curvature of streamlines near the body are situated near the bottom of the page 41.) On the other hand, this instructive indication of why lift occurs leads to no practical method for calculating it.

4
Vorticity fields

We have seen that the solenoidality property (eqn (42) giving constancy of Sq along a streamtube) and, in steady frictionless flow, the Bernoulli property (eqn (86): constancy of $p_e + \frac{1}{2}\rho q^2$ along a streamline) are important simplifications which throw valuable light on the motions of fluids. Nevertheless, because they are only two relationships, they cannot give full quantitative information regarding three-dimensional motions specified in general by four unknown quantities[†] p_e, u, v, and w. It was, accordingly, only after another brilliant simplification had been introduced into theoretical fluid mechanics, and developed through the work of Joseph Lagrange (1736–1813), Augustin Cauchy (1789–1857), Hermann von Helmholtz 1821–94), George Stokes (1819–1903), and Lord Kelvin (1824–1907), that significant progress could be made in the analysis of the forbiddingly nonlinear equations of motion (eqns (69)).

As a result of their work, we recognize that the essential additional key needed for simplifying the study of fluid motions lies in the related concepts of vorticity and circulation, which are developed in the present chapter and which are continually applied throughout the rest of this book. These concepts are founded on an acceptance of the need to analyse the *rotation* of a particle of fluid.

Thus, just as it is important in the mechanics of a rigid body to study its rotation and the associated *angular momentum*, so also these properties can valuably be studied for a particle of fluid. The fluid particle, however, exhibits a major difference from a rigid body: it does not simply rotate (in its motion relative to its mass centre) but is also deformed at the same time in ways which we analyse in some detail below.

4.1 Analysis of instantaneous deformation of a particle

In Sections 4.1 to 4.3 we make a careful study of the instantaneous rate of deformation which a very small *spherical* particle of fluid undergoes in the velocity field of expression (35). Consider, then, a particle of fluid P in the form of a sphere of very small radius ε whose geometrical centre (which is also its mass centre, since the fluid is assumed to be of constant density) is a point C with coordinates (x, y, z).

The sphere is assumed to be so small that the coordinates of any point within

---

[†] It is, indeed, only in an essentially one-dimensional motion like that described in Section 1.2, with just two unknown quantities, that those two relationships suffice.

it can be written

$$(x + dx, y + dy, z + dz) \tag{88}$$

in the sense defined in the footnote on p. 14. Thus, the expressions

$$(dx, dy, dz) \tag{89}$$

represent the position of such a general point within the particle of fluid *relative to* the position C of its mass centre.

Similarly, we write the fluid velocity at the general point in expression (88) in the form

$$(u + du, v + dv, w + dw) \tag{90}$$

where (u, v, w) stands for the fluid velocity at the mass centre C. Then the expressions

$$(du, dv, dw) \tag{91}$$

represent the velocity of any point within the particle of fluid relative to that of its mass centre C.

The definition of a partial differential coefficient tells us that, for the velocity field of expression (35), we can write the instantaneous values of the velocity components in expression (91) (that is, their values at a fixed time t) as follows:

$$\left. \begin{array}{l} du = (\partial u/\partial x)dx + (\partial u/\partial y)dy + (\partial u/\partial z)dz, \\ dv = (\partial v/\partial x)dx + (\partial v/\partial y)dy + (\partial v/\partial z)dz, \\ dw = (\partial w/\partial x)dx + (\partial w/\partial y)dy + (\partial w/\partial z)dz. \end{array} \right\} \tag{92}$$

Note that, in the eqns (92), there is no term in dt (representing a change in the time t) because the components are being considered at a fixed time t.

In the matrix notation with which most readers will be familiar,[†] it is sometimes convenient to abbreviate the eqns (92) to the form

$$(du, dv, dw) = A(dx, dy, dz)^T. \tag{93}$$

Here, A is the 3×3 square matrix whose elements are the bracketed expressions in eqns (92), while T stands for transpose, so that its application to the row vector (dx, dy, dz) signifies the corresponding column vector.

Before proceeding, in Section 3.2, to study the rotation and angular momentum of the particle P, consider first its linear momentum. We can write this as

$$\int_P (u + du, v + dv, w + dw)\rho dV, \tag{94}$$

an integral of the velocity in expression (90) times an element ρdV of the particle's mass over the whole extent of the spherical volume of radius ε

[†] In fact, only at a few places in the present book (mainly in Section 4.3) are results expressed in matrix form. One alternative which might have been adopted (the use of tensor concepts) is, on the other hand, avoided in order to keep the text as widely intelligible as possible.

specified by the inequality.

$$(dx)^2 + (dy)^2 + (dz)^2 \leqslant \varepsilon^2. \tag{95}$$

In the integral expression (94), the expressions du, dv, dw are those given by eqns (92), and their integrals over the spherical volume defined by eqn (95) are all zero. (Mathematically, the integral of, say, dx over the volume is zero because the volume is completely symmetrical about the plane $dx = 0$. Physically, this is so because the mean of the *displacement from the mass centre* (expression (89)) is necessarily zero. The same applies to the integral of dy or dz.) Therefore, expression (94) reduces to

$$m(u, v, w), \tag{96}$$

where the integral of the constant density ρ over P is

$$m = \tfrac{4}{3}\pi\varepsilon^3\rho, \tag{97}$$

the mass of the particle. In words, the momentum of the particle P is the same as if the whole particle were moving with the velocity (u, v, w) of its mass centre. This, we may note, is one general property of rigid bodies which P happens to share even though it is not rigid.

4.2 Vorticity

For any mechanical system, a measure of the system's *rotation* is given by its angular momentum about the mass centre. That angular momentum is a vector whose component in any direction may be described as the angular momentum, or 'moment of momentum', about an axis through the mass centre pointing in that direction.

In particular, the x component of angular momentum of the particle P about C is the moment of momentum about an axis through C pointing in the x direction. This can be written in the form of an integral as in expression (94), but with the integrand replaced by its moment about that axis:

$$\int_P [dy(w + dw) - dz(v + dv)]\rho dV, \tag{98}$$

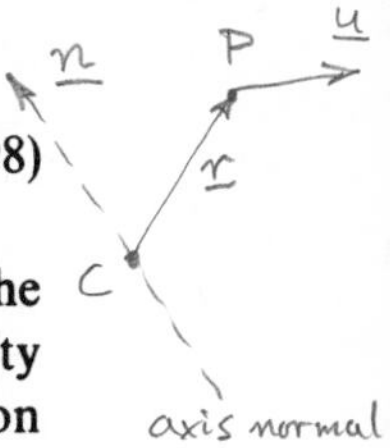

where the expression in square brackets represents the x-component of the vector product of the displacement vector (expression (89)) with the velocity vector (expression (90)) (that is the x-component of the moment of expression (90) about P).

The integral in expression (98) can be rewritten, using eqns (92), as

$$\int_P dy[(\partial w/\partial x)dx + (\partial w/\partial y)dy + (\partial w/\partial z)dz]\rho dV$$

$$- \int_P dz[(\partial v/\partial x)dx + (\partial v/\partial y)dy + (\partial v/\partial z)dz]\rho dV. \tag{99}$$

[handwritten margin notes:]

n P u

r

C

axis normal to plane of r & u

Angular momentum of particle at P, a (per unit mass):

$a = r \wedge u.$

(Easily seen from definition of $A \wedge B = |A| |B| \sin\theta \; n$.)

Here, the contributions from dw and dv in expression (98) have been written out at length. However, the contributions from w and v have been recognized as zero and suppressed. (We saw in Section 4.1 that the integral of either dy or dz over P is exactly zero.). In words, only the velocities in expression (91) *relative* to that of C contribute to the particle's angular momentum about C.

In the first line of expression (99) the integrals of two of the three terms are zero and, similarly, two in the second line are zero; in each case this is because

$$0 = \int_P (dx\,dy)\rho dV = \int_P (dy\,dz)\rho dV = \int_P (dz\,dx)\rho dV. \tag{100}$$

Mathematically, the integrals in eqn (100) over the spherical volume defined in eqn (95) must vanish because that volume is symmetrical about each of the planes d$x = 0$, d$y = 0$, and d$z = 0$. Physically, they represent the particle's *products of inertia*, which are always zero for a homogeneous sphere.

On the other hand, we may evaluate the non-zero term on either line of expression (99) in terms of the integral expressions

$$\int_P (dy)^2 \rho dV = \int_P (dz)^2 \rho dV = \tfrac{1}{2}I, \tag{101}$$

where

$$I = \int_P [(dy)^2 + (dz)^2]\rho dV = \tfrac{8}{15}\rho\varepsilon^5 = \tfrac{2}{5}m\varepsilon^2 \tag{102}$$

is the familiar expression for the *moment of inertia* of a homogeneous sphere P about any axis through its centre C. The expression (99) for the x component of angular momentum is seen, using eqns (100) and (101), to take the reasonably compact form

$$\tfrac{1}{2}I(\partial w/\partial y - \partial v/\partial z). \tag{103}$$

Similar evaluation of the y and z components leads to an expression for the complete angular momentum vector as

$$\tfrac{1}{2}I(\partial w/\partial y - \partial v/\partial z, \ \partial u/\partial z - \partial w/\partial x, \ \partial v/\partial x - \partial u/\partial y). \tag{104}$$

Indeed, the need for the y and z components to take the forms shown in expression (104) would be evident without further calculation since they are derived by cyclically permuting x, y, z, and u, v, w.

For any velocity field (expression (35)), the associated vector field given by the bracketed part of expression (104) is called the *vorticity* field. We have seen that its value at a point C specifies the angular momentum (expression (104)) of a small sphere centred on C; thus it is a measure of local rotation.

Many readers will, indeed, be aware that for any vector field **u** whatsoever, which has components (u, v, w), the vector field within brackets in expression

(104) is the field widely known as the 'curl' of **u** and written in either of the forms

$$\text{curl } \mathbf{u} = \nabla \times \mathbf{u} = (\partial w/\partial y - \partial v/\partial z, \partial u/\partial z - \partial w/\partial x, \partial v/\partial x - \partial u/\partial y). \quad (105)$$

Note that the second expression in eqn (105) provides a useful way of remembering the components of curl **u**, which it views as a vector product of the operator

$$\nabla = (\partial/\partial x, \partial/\partial y, \partial/\partial z) \quad (106)$$

(the gradient operator) with **u**, just as the divergence (*see* eqn (45)) can be viewed as the scalar product of eqn (106) with **u**. The curl as defined in eqn (105) plays an important role in many classical theories, such as electromagnetic theory.

In theoretical fluid mechanics the letter used to denote the vorticity is the same Greek omega,

$$\boldsymbol{\omega} = \text{curl } \mathbf{u}, \quad (107)$$

which in rigid-body mechanics is commonly used to signify angular velocity. It is necessary to warn the reader, however, that this notation involves two dangers of misunderstanding. First, the particle P is *not* a rigid body (indeed, its instantaneous rate of deformation from the spherical shape will be determined in Section 4.3); and second, if it *were* a rigid sphere with moment of inertia I, then its angular momentum, given by expression (104), and eqns (105) and (107), as

$$\tfrac{1}{2}I\boldsymbol{\omega}, \quad (108)$$

would imply that its angular velocity must take the value $\tfrac{1}{2}\boldsymbol{\omega}$ rather than $\boldsymbol{\omega}$. In short, the vorticity $\boldsymbol{\omega}$ at the centre of P identifies the angular momentum of P as that which P would possess if it were rigidly rotating with angular velocity, not $\boldsymbol{\omega}$, but $\tfrac{1}{2}\boldsymbol{\omega}$.

4.3 Rate of strain

The remarks just made significantly help us towards the determination of the *rate of strain* (rate of change of shape) of the particle P. To this end we consider the relative velocity components in expression (91) as specified in eqn (93) and compare them with those found in a motion without any change of shape but with identical angular momentum, that is in a rigid-body rotation about C with angular velocity $\tfrac{1}{2}\boldsymbol{\omega}$.

In such a rigid-body rotation the relative velocities would be given by the vector product of the angular velocity with the position vector (expression (89)) relative to the centre of rotation

$$(du, dv, dw) = (\tfrac{1}{2}\boldsymbol{\omega}) \times (dx, dy, dz)$$
$$= \tfrac{1}{2}(\eta dz - \zeta dy, \zeta dx - \xi dz, \xi dy - \eta dx), \quad (109)$$

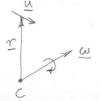

Solid body rotation, angular velocity $\underline{\omega}$:

$$|\underline{u}| = |\underline{\omega}||\underline{r}|$$
$$\hat{\underline{u}} = \hat{\underline{\omega}} \wedge \hat{\underline{r}}$$
$$\Rightarrow \underline{u} = \underline{\omega} \wedge \underline{r}$$

where the components of the vector product have been written out in terms of the components   xi , eta , zeta

$$(\xi, \eta, \zeta) \tag{110}$$

of the vorticity $\boldsymbol{\omega} = \operatorname{curl} \mathbf{u}$.

To compare the rigid-body motion of eqn (109) with the true motion of the particle P as spcified in eqn (93), we rewrite eqn (109) as

$$(du, dv, dw) = B(dx, dy, dz)^{\mathrm{T}}, \tag{111}$$

with the matrix B specified as

$$B = \tfrac{1}{2} \begin{pmatrix} 0 & -\zeta & \eta \\ \zeta & 0 & -\xi \\ -\eta & \xi & 0 \end{pmatrix}. \tag{112}$$

Here, the values of ξ, η, and ζ (the components in expression (110) of curl $\mathbf{u}$) are as noted in eqn (105). When they are substituted in eqn (112), and the resulting matrix B is compared with the matrix A of eqn (93) whose elements are the bracketed expressions in eqns (92), we observe the interesting fact that

$$B = \tfrac{1}{2}(A - A^{\mathrm{T}}). \tag{113}$$

Here, the transpose A^{T} of the square matrix A has the same diagonal elements as A, which is why the difference $A - A^{\mathrm{T}}$ has zero diagonal elements as in eqn (112), while its other elements are the mirror images of the elements of A, reflected in that main diagonal. This is why the difference $A - A^{\mathrm{T}}$ is an antisymmetric matrix taking the form of eqn (112).

This remarkable result (eqn (113)) allows us to write the true motion (eqn (93)) of the fluid particle as a simple sum of (i) a rigid-body motion (eqn (111)) with the same angular momentum as P, and (ii) a pure rate of strain (that is, a rate of change of *shape* with zero angular momentum) given by

$$(du, dv, dw) = E(dx, dy, dz)^{\mathrm{T}}. \tag{114}$$

Here, E (the difference of A and B) is the symmetric matrix

$$E = \tfrac{1}{2}(A + A^{\mathrm{T}}), \tag{115}$$

so that the decomposition used is equivalent to the standard procedure in algebra by which any matrix A is expressed as the sum

$$A = \tfrac{1}{2}(A + A^{\mathrm{T}}) + \tfrac{1}{2}(A - A^{\mathrm{T}}) = E + B \tag{116}$$

of a symmetric matrix E and an antisymmetric matrix B.

The fact that a pure rate of strain as in eqn (114) with zero angular momentum is specified by a symmetric matrix E allows us to visualise the nature of that motion (*see* eqn (114)) with the aid of one more standard result from matrix theory. This states that, for any symmetric matrix E, we can rotate the axes used (that is, we can choose a new system of Cartesian coordinates x', y', z' which are related to x, y, z by an orthogonal matrix L) so that the equations corresponding to eqn (114) in the new axes are

$$du' = \alpha dx', \quad dv' = \beta dy', \quad dw' = \gamma dz', \tag{117}$$

the constants α, β, and γ being *eigenvalues* of E. Here, the components in the old and new axes are related by equations

$$(x', y', z') = L(x, y, z)^{\mathrm{T}}, \quad (u', v', w') = L(u, v, w)^{\mathrm{T}}, \tag{118}$$

so that eqn (114) gives

$$(du', dv', dw') = LE(dx, dy, dz)^{\mathrm{T}} = LEL^{-1}(dx', dy', dz'). \tag{119}$$

The standard result used states, in fact, that an orthogonal matrix L (i.e. one such that $L^{-1} = L^{\mathrm{T}}$, so that the coordinate system remains Cartesian) can always be found so that the transformed symmetric matrix LEL^{-1} takes the diagonal form

$$LEL^{-1} = \begin{pmatrix} \alpha & 0 & 0 \\ 0 & \beta & 0 \\ 0 & 0 & \gamma \end{pmatrix}, \tag{120}$$

which reduces eqn (119) to the simple set of eqns (117).

Figure 20 illustrates both the change of axes (eqn (118)) and the nature of the pure rate of strain motion (eqns (117)) in the new axes. A case where $\alpha > 0$ is illustrated so that the first of the eqns (117) describes a uniform rate of *elongation* in the x' direction; on the other hand, $\beta < 0$ so that the second equation describes a uniform rate of *foreshortening* in the y' direction (the figure cannot, of course, display deformation in the z' direction perpendicular to the paper). During a short time dt, these processes are deforming the sphere of radius ε into an ellipsoid whose semiaxes have lengths

$$\varepsilon(1 + \alpha dt), \quad \varepsilon(1 + \beta dt), \quad \varepsilon(1 + \gamma dt), \tag{121}$$

since, for example, the eqns (117) make $u' = \varepsilon \alpha$ and $v' = w' = 0$ at the end of the first semiaxis, where $x' = \varepsilon$ and $y' = z' = 0$. The axes of the ellipsoid in Fig. 20 (pointing in the x', y', and z' directions) are usually called the *principal axes* of rate of strain.

The rate of change of volume of the ellipsoid is readily calculated to be zero because the equation of continuity (eqn (42)) in the new axes gives

$$\alpha + \beta + \gamma = \partial u'/\partial x' + \partial v'/\partial y' + \partial w'/\partial z' = 0. \tag{122}$$

This is why, if one of α, β, or γ is positive (indicating an elongation) then, as in Fig. 20, at least one other must be negative (indicating a foreshortening).

Besides the pure rate of strain which is analysed in this section, Fig. 20 also illustrates the other two constituents of the motion of the particle P. These are:

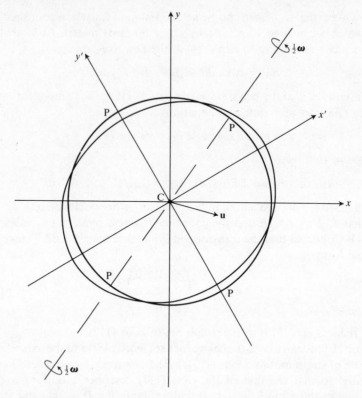

Fig. 20. The instantaneous rate of deformation of a small spherical particle of fluid P is analysed into three constituents: (i) uniform translation at the velocity **u** of the particle's centre C; (ii) rigid rotation with angular velocity $\frac{1}{2}$ **ω** where **ω** is the vorticity at C; and (iii) pure rate of strain. This last constituent, after a particular change of axes (eqns (118)) from (x, y, z) to (x', y', z') (the principal axes of rate of strain), takes a simplified form; say, a uniform elongation in the x' direction, together with a uniform foreshortening in the y' direction and either an elongation or a foreshortening in the z' direction. All of the particle's momentum is in constituent (i), all of its angular momentum is in constituent (ii), and all of its rate of change of shape is in constituent (iii).

the uniform translation at a uniform velocity (the velocity (u, v, w) of the mass centre C relative to which all the components in expression (91) are measured); and the rigid rotation at an angular velocity $\frac{1}{2}$**ω**. These were shown in Sections 4.1 and 4.2 to embody all of the particle's momentum, and all of its angular momentum, respectively. The third constituent (pure rate of strain) is without momentum or angular momentum, but embodies all of the particle's rate of change of shape.

Figure 21 illustrates a particular motion, often described as a shearing motion, which is of importance to the study of boundary layers (Sections 3.4

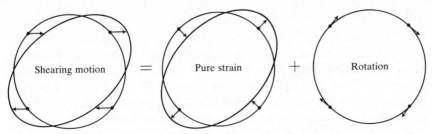

Fig. 21. The shearing motion (expression (123)) has the entire fluid moving in a single direction (the x direction), at velocities varying linearly with the y coordinate. Yet the analysis of even such a unidirectional motion into its constituents as described in Fig. 20 yields a *rotational* constituent as well as pure rate of strain. The rotation is at angular velocity $\frac{1}{2}\boldsymbol{\omega}$, where the magnitude of the vorticity vector $\boldsymbol{\omega}$ is necessarily equal to the rate of shear s.

and 5.5). The velocity field possesses only an x component but its magnitude increases linearly with y, thus

$$\mathbf{u} = (sy, 0, 0) \text{ so that } \boldsymbol{\omega} = (0, 0, -s), \tag{123}$$

where the 'rate of shear' s is constant. The matrix A of eqn (93) has the properties

$$A = \begin{pmatrix} 0 & s & 0 \\ 0 & 0 & 0 \\ 0 & 0 & 0 \end{pmatrix}, \quad E = \tfrac{1}{2}(A + A^{\mathrm{T}}) = \begin{pmatrix} 0 & \tfrac{1}{2}s & 0 \\ \tfrac{1}{2}s & 0 & 0 \\ 0 & 0 & 0 \end{pmatrix}, \tag{124}$$

and E is reduced to diagonal form by a rotation of axes through an angle $\frac{1}{4}\pi$, which is represented by eqn (118) with

$$L = 2^{-\frac{1}{2}} \begin{pmatrix} 1 & 1 & 0 \\ -1 & 1 & 0 \\ 0 & 0 & 2^{\frac{1}{2}} \end{pmatrix}, \quad LEL^{-1} = \begin{pmatrix} \tfrac{1}{2}s & 0 & 0 \\ 0 & -\tfrac{1}{2}s & 0 \\ 0 & 0 & 0 \end{pmatrix}. \tag{125}$$

In time dt, then, the pure rate of strain E changes the spherical particle P into an ellipsoid of semiaxes

$$\varepsilon(1 + \tfrac{1}{2}s\,dt), \quad \varepsilon(1 - \tfrac{1}{2}s\,dt), \varepsilon, \tag{126}$$

where the first, say, is inclined at an angle $\frac{1}{4}\pi$ to the x axis. However, simultaneous rotation with angular velocity $\frac{1}{2}\boldsymbol{\omega}$ is seen by eqn (123) to produce a clockwise turning which slightly reduces that angle to $\frac{1}{4}\pi - (\tfrac{1}{2}s)\,dt$, and Fig. 21 illustrates well how, in order to represent the shearing motion accurately, it is essential that this rotation accompanies the rate of strain. Vorticity, in short, is a crucially important feature of any shearing motion.

4.4 Vortextubes

Of the two concepts introduced in Sections 4.2 and 4.3, it is indeed the vorticity concept which is primarily used throughout the rest of this book. Relatively, the rate-of-strain concept is rarely used, but it has been necessary to explain it in detail so that the vitally important concept of vorticity could be fully understood as a measure of rotation of a fluid particle which, in general, is accompanied by simultaneous elongations and foreshortenings.

It has been pointed out in Section 4.2 that for any velocity field **u** the vorticity curl **u** is itself a vector field whose components are given in eqn (105). The special properties of *the vorticity field* **ω** are now considered.

First of all we note that the vorticity field is *solenoidal* (div **ω** = 0). Indeed, for any vector field **u** we have

$$\text{div curl } \mathbf{u} = 0. \tag{127}$$

This is easily checked by applying the divergence operator of eqn (45) to the curl **u** field defined in eqn (105). Another way of writing eqn (127),

$$\nabla \cdot (\nabla \times \mathbf{u}) = 0, \tag{128}$$

makes the result seem very natural as the vanishing of a 'scalar triple product' in which the gradient operator ∇ appears twice.

From Section 2.4, furthermore, we know what the solenoidality property means for a vector field's *geometry*. This is related to the properties of a pattern of lines generated as follows: draw large numbers of very short arrows, each proportional to the local value of the vector field, and join those arrows up to form an assemblage of curves. In Fig. 13 this was done for the velocity field in order to form streamlines which, mathematically, are specified by the eqns (37) whether in steady or unsteady flow. Later (Fig. 14), we defined a streamtube as composed of the streamlines passing through a particular loop. Finally, we considered a *thin* streamtube (Fig. 15), and showed two forms of the solenoidality property to be equivalent: the constancy of Sq (streamtube area times the magnitude of the vector field) along such a thin streamtube, and the vanishing of the divergence.

Proceeding by analogy (Fig. 22), we can draw large numbers of very short arrows, each proportional to the local value of the vorticity field, and join these up to form an assemblage of curves called vortexlines. Mathematically, they are specified (at each time t) by the equations

$$\frac{\mathrm{d}x}{\xi} = \frac{\mathrm{d}y}{\eta} = \frac{\mathrm{d}z}{\zeta}. \tag{129}$$

Next, (Fig. 23) we draw (i) a general vortextube composed of the vortexlines passing though a particular loop, and (ii) a very thin vortextube. For either (i) or (ii), the solenoidality property div **ω** = 0 implies that all along the

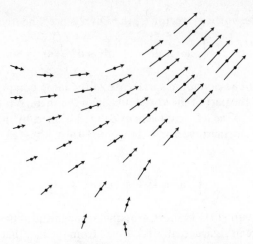

Fig. 22. Just as velocity vectors can be joined up to form streamlines (Fig. 13), so vorticity vectors can be joined up to form vortexlines.

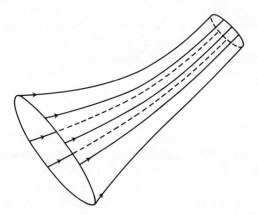

Fig. 23. A vortextube is a surface comprising all the vortexlines passing through points of a particular loop.

vortextube a certain quantity takes an unchanging value. For the thin vortextube (ii) that quantity is expected to be the product of its cross-sectional area with the *magnitude* of the vorticity (and this is verified below). Wherever the tube narrows greatly, then, the fluid's local rotation must be greatly enhanced.

For a general vortextube (i), we obtain a similar result by applying the Divergence Theorem (eqn (48)) to a particular lump L of fluid inside the vortextube. This is the lump bounded by two different surfaces S_1 and S_2 (Fig. 23) spanning the tube and by the part of the tube between them.

Applied to the vorticity vector $\boldsymbol{\omega}$, the Divergence Theorem gives

$$\int_{\partial L} \boldsymbol{\omega} \cdot \mathbf{n} \, dS = \int_{L} \text{div} \, \boldsymbol{\omega} \, dV = 0, \tag{130}$$

since div $\boldsymbol{\omega} = 0$. The bounding surface ∂L of the lump comprises not only S_1 and S_2 but also the part of the vortextube between them; but the latter makes no contribution to the left-hand side of eqn (130) since, by the definition of a vortextube, the vorticity vector is tangential to it, thus giving $\boldsymbol{\omega} \cdot \mathbf{n} = 0$. It follows that

$$\int_{S_1} \boldsymbol{\omega} \cdot \mathbf{n} \, dS + \int_{S_2} \boldsymbol{\omega} \cdot \mathbf{n} \, dS = 0. \tag{131}$$

The integrals in eqn (131), in short, are equal in magnitude but opposite in sign.

The difference in sign, actually, is derived from the fact that the Divergence Theorem (eqn (48)) takes $\mathbf{n}$ as a unit vector along the *outward* normal from the lump L. Thus, as Fig. 23 suggests, the second integral in eqn (131) is positive ($\mathbf{n}$ has a positive component in the direction of $\boldsymbol{\omega}$) but the first is negative (as is the corresponding component).

If, however, the direction of the normal $\mathbf{n}$ is changed in the first integral of eqn (131), but not in the second, by redefining $\mathbf{n}$ in integrals over surfaces spanning the vortextube so that it always has a positive component in the direction of $\boldsymbol{\omega}$, then we can express eqn (131) in the simpler form

$$\int_{S} \boldsymbol{\omega} \cdot \mathbf{n} \, dS = \text{constant} \tag{132}$$

for all surfaces S spanning a given vortextube.

Application of eqn (132) to a simple cross-section S of a *thin* vortextube gives the useful result already noted (the product of S with the magnitude of the vorticity remains constant along it). However, the more general result of eqn (132) for any vortextube is found also to be extremely valuable, for reasons which we now probe further.

4.5 Circulation

The quantity stated in eqn (132) to take everywhere a constant value along a general vortextube is a characteristic property of that vortextube, sometimes called its 'strength'. We consider now whether any special physical interpretation can be assigned to this quantity. For example, the corresponding quantity for a general *streamtube* (Fig. 14) would be the integral

$$\int_{S} \mathbf{u} \cdot \mathbf{n} \, dS \tag{133}$$

across a surface spanning the streamtube. The physical interpretation of this quantity is readily seen to be the rate of *volume flow* along the streamtube.

It is a famous mathematical theorem due to Stokes which gives a special interpretation to the left-hand side of eqn (132). Stokes's theorem applies to any surface S for which we are able to make a consistent, continuously varying selection of a direction for the unit normal vector $\mathbf{n}$ at each point of the surface[†] and for which the boundary ∂S consists of one or more closed curves (Fig. 24). Stokes's theorem states that, for any vector field $\mathbf{u}$,

$$\int_S (\operatorname{curl} \mathbf{u}) \cdot \mathbf{n} \, dS = \int_{\partial S} \mathbf{u} \cdot d\mathbf{x}. \tag{134}$$

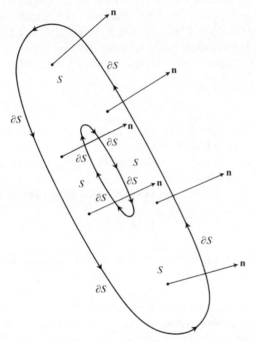

Fig. 24. Illustrating Stokes's theorem (eqn (134)) for a surface S whose boundary ∂S consists of two closed curves. The integral on the right-hand side of eqn (134) is taken around each curve in the direction specified by arrows, namely in the positive sense relative to the direction chosen for the normals $\mathbf{n}$ to the surface S.

[†] For any ordinary surface with two sides to it this simply means choosing one of those two sides and having $\mathbf{n}$ point everywhere outwards from *that* side rather than from the other. It is only to peculiar surfaces, e.g. Möbius bands which do not possess two sides, that Stokes's theorem does not apply.

Here, the integral of

$$\mathbf{u} \cdot d\mathbf{x} = u dx + v dy + w dz \tag{135}$$

along each of the closed curves making up the boundary ∂S must be taken 'in the positive sense' relative to the direction chosen for $\mathbf{n}$. (Accordingly, if the opposite direction for the normal $\mathbf{n}$ were used, both sides of eqn (134) would change sign.)

The reader may, perhaps, like to be reminded that 'the positive sense' relative to $\mathbf{n}$ means the sense in which a screwdriver would turn about $\mathbf{n}$ to drive a right-handed screw in the direction of $\mathbf{n}$. Alternatively, if a person's right hand is held all in one plane, with the thumb at right angles to the other fingers, the positive sense relative to the direction of the rigidly held thumb is the sense in which the tips of the other fingers must move in order to close the fist.

In the rest of this section we first sketch, for the case when $\mathbf{u}$ is a velocity field, a little of the significance of Stokes's theorem (with which many readers are, perhaps, already familiar). The section ends, however, with a brief derivation of the theorem to help readers meeting it for the first time.

For any closed curve C, the integral of the velocity field,

$$\int_C \mathbf{u} \cdot d\mathbf{x} \tag{136}$$

taken around C in a particular sense, is given the special name *circulation*. More precisely, it is called the circulation in that sense around C. Here, the magnitude of the small vector displacement $d\mathbf{x}$ along C is a small element ds of arc length along C, while its direction is that of $\mathbf{t}$, a unit *tangential* vector in the chosen sense. Thus, the circulation (expression (136)) can be written

$$\int_C (\mathbf{u} \cdot \mathbf{t}) ds; \tag{137}$$

and, for example, it takes a positive value if, on an average with respect to arc length along the curve, positive values of the tangential resultant $\mathbf{u} \cdot \mathbf{t}$ of the velocity field outweigh negative values.

There are some special *dynamical* reasons (*see* Section 5.1) why circulation is important, but we confine ourselves here to its importance for the kinematics of vortextubes. For any surface S spanning a vortextube (Fig. 23), its boundary is a single closed curve C embracing the vortextube once, and any such curve C is the boundary of a spanning surface S. Stokes's theorem (eqn (134)) tells us, then, that

$$\int_S \boldsymbol{\omega} \cdot \mathbf{n} \, dS = \int_C \mathbf{u} \cdot d\mathbf{x}; \tag{138}$$

that is, the 'strength' of the vortextube (the quantity which takes the same constant value (eqn (132)) for any surface spanning it) is equal to the circulation around any closed curve C which embraces it once, taken in the positive sense with respect to the direction of the normal **n**.

The simple approach to a derivation of Stokes's theorem with which this chapter is concluded shows first that for a *flat* surface S (part of a plane) eqn (134) follows immediately from the Divergence Theorem (Section 2.4). This approach then deduces it for a general surface by approximating that surface with small flat (e.g. triangular) elements.

To facilitate the proof of eqn (134) for a flat surface S we are free to adopt coordinates x, y, and z for which S is part of the plane $z = 0$ and **n** is a unit vector in the z direction. Then eqn (134) becomes

$$\int_S (\partial v/\partial x - \partial u/\partial y)\,\mathrm{d}S = \int_{\partial S} (ut_x + vt_y)\,\mathrm{d}s \tag{139}$$

in terms of the z component of eqn (105) and the components $(t_x, t_y, 0)$ of the unit tangential vector which appears in the form of expression (137) of the right-hand side of eqn (134).

If, however, we apply the Divergence Theorem (eqn (48)) to a lump L in the form of a wafer-thin slice extending between $z = -\tfrac{1}{2}\varepsilon$ and $z = +\tfrac{1}{2}\varepsilon$ on both sides of S, and to the vector field with components $(v, -u, 0)$, we obtain eqn (139) multiplied by the slice's very small thickness ε. This is because: (i) the divergence of the vector field $(v, -u, 0)$ is the integrand on the left-hand side of eqn (139); (ii) this vector field has zero z component, that is zero normal component on the flat faces $z = \pm\tfrac{1}{2}\varepsilon$ of ∂L which therefore make no contribution to the right-hand side of eqn (48); and (iii) on the boundary ∂S the outward normal component of $(v, -u, 0)$ is the integrand on the right-hand side of eqn (139), namely (in Fig. 25) the *tangential* component of a vector

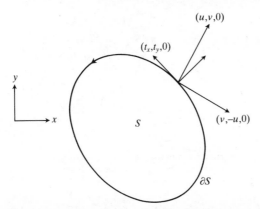

Fig. 25. Stokes's theorem is here applied to a flat surface S with the normals to S pointing out of the paper towards the reader. A component $ut_x + vt_y$ of $(u, v, 0)$ along the tangent to ∂S is identical with a component of the perpendicular vector $(v, -u, 0)$ along the outward normal to ∂S.

$(u, v, 0)$ obtained by rotating $(v, -u, 0)$ through $90°$ about the z axis in the positive sense.

This establishes Stokes's theorem for any flat surface. Finally, a surface S of general shape can be successively approximated by large numbers of finite flat elements; for example, in the process known as triangulation (Fig. 26). We may then add up the results of applying Stokes's theorem to each of the triangles. Provided that the normal **n** has everywhere been taken pointing outwards from the same *side* of S (see the footnote on p. 55), then it is the contributions from ∂S (more correctly, from edges of triangles representing ∂S) which appear just once in, and remain after, such an addition. By contrast, contributions from edges of triangles inside S are taken twice, in opposing senses (Fig. 26), and therefore cancel out. This argument proves Stokes's theorem for every one of the triangulated approximations to S, whence its truth for S itself (their limiting form) follows.

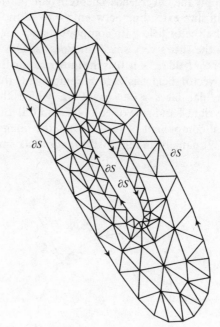

Fig. 26.   The process of triangulation used for approximating to a general surface S by large numbers of finite flat elements.

5
Vortex dynamics

The meanings, and interrelationships, of vorticity and circulation have been developed in Chapter 4 from the pure standpoint of 'kinematics' (analysis of motions without considering the 'dynamical' effect of forces). The present chapter begins the study of their very fruitful contribution to an understanding of the dynamics of fluid motions.

A single discovery by Kelvin (his remarkable 1869 theorem on persistence of circulation) is used to derive all the theoretical results given in this chapter, including some discoveries which, historically, had been obtained otherwise, and earlier, by Lagrange, Cauchy and Helmholtz. (The same circulation theorem finds further important applications in Chapters 10 and 11.) The chapter ends with a first attempt at giving insight into the properties of boundary layers, based on a combination of theoretical understanding and experimental data.

5.1 The persistence of circulation

The persistence ascribed to circulation by Kelvin's theorem does not refer to the circulation around a curve C whose shape and position are fixed in space. The theorem is concerned, rather, with a closed curve C consisting always of the same particles of fluid. We should think of C, then, as a *necklace of particles of fluid*, one whose shape and position change continually as those particles of fluid move. This is why it is usually described as 'a closed curve moving with the fluid'.

Kelvin's theorem is exact for the Euler model, being derived (as we shall see) from the form (eqn (71)) of the momentum principle in that model. It states that *the circulation around a closed curve C moving with the fluid remains constant*.

It is worth noting in the following proof of Kelvin's theorem that the assumptions of the Euler model, of which the most restrictive was seen in Section 2.1 to be the neglect of viscous forces, are used only at points along C itself. This means that, besides being exact for the Euler model, the persistence of circulation around a curve moving with the fluid is a very close approximation for any curve which avoids regions (such as boundary layers) where viscous forces are important.

Consider now a curve C which, because it is made up of a necklace of identified particles of fluid, can be described as moving with the fluid. Assuming the momentum equation (eqn (71)) at each point of C, we deduce

Kelvin's result that the rate of change of the circulation (expression (136)) around C is zero.

To this end, consider two adjacent particles of fluid in the necklace C (Fig. 27) whose position vectors are x and $x + dx$. It is their *relative* position dx which forms a scalar product with the velocity u in the circulation

$$\int_C u \cdot dx. \tag{140}$$

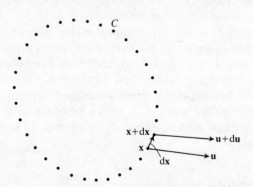

Fig. 27. Kelvin's theorem is concerned with the rate of change of the circulation (expression (140)) around a closed curve C consisting always of the same particles of fluid. Here dx, the vector separation between two nearby particles on C, is changing at a rate du, their *relative* velocity.

As both particles move, that relative position dx is changing at a rate equal to the relative *velocity* of the two particles. Taking the fluid velocity as u at the point x and as $u + du$ at the point $x + dx$ (somewhat as in Section 4.1), the relative velocity of the two particles is du. In short, just as an application of the operator D/Dt (the rate of change (eqn (62)) following a particle) to the position $x = (x, y, z)$ itself gives the velocity u, so the rate of change of the relative position dx of two particles, following the particles, is their relative velocity:

$$D(dx)/Dt = du. \tag{141}$$

Now, using the general form for the rate of change of a scalar product, we can write down the rate of change of the scalar product in expression (140) following the particles as

$$D(u \cdot dx)/Dt = u \cdot D(dx)/Dt + (Du/Dt) \cdot dx, \tag{142}$$

but the actual circulation is an integral (expression (140)) around the closed curve C; that is, the limit of a sum of elements such as $u \cdot dx$, whose rate of change is given in eqn (142). The rate of change of the expression (140) for a necklace C of particles each moving with the fluid is therefore just such an integral of the right-hand side of eqn (142):

$$D\left(\int_C \mathbf{u} \cdot d\mathbf{x}\right)/Dt = \int_C \mathbf{u} \cdot d\mathbf{u} + \int_C (D\mathbf{u}/Dt) \cdot d\mathbf{x}, \qquad (143)$$

where eqn (142) has been simplified by using eqn (141).

Both terms on the right-hand side of eqn (143) are, however, quickly seen to be zero. The first can be written, using the rule for the differential of a product, as the integral

$$\int_C d(\tfrac{1}{2}\mathbf{u} \cdot \mathbf{u}) \qquad (144)$$

around the closed curve C. This gives the difference between the values of $\tfrac{1}{2}\mathbf{u} \cdot \mathbf{u}$ at the end points of the path of integration, a difference which is evidently zero since, for a closed curve, the end points coincide. The second can be written, using the momentum equation[†] (eqn (71)), as

$$-\rho^{-1}\int_C (\text{grad } p_e) \cdot d\mathbf{x} = -\rho^{-1}\int dp_e = 0, \qquad (145)$$

because

$$(\text{grad } p_e) \cdot d\mathbf{x} = (\partial p_e/\partial x)dx + (\partial p_e/\partial y)dy + (\partial p_e/\partial z)dz \qquad (146)$$

is dp_e (the difference in p_e between adjacent particles on the curve), and because, as before, the integral of dp_e (the sum of all those differences around the closed curve) is zero. Finally, then, the equation

$$D\left(\int_C \mathbf{u} \cdot d\mathbf{x}\right)\bigg/Dt = 0 \qquad (147)$$

describing the persistence of circulation around a closed curve moving with the fluid, has been established. Applications of this theorem of Kelvin abound in the rest of this chapter and in Chapters 10 and 11.

5.2 The movement of vortexlines

One especially valuable deduction from Kelvin's theorem is concerned with 'the movement of vortexlines'. This is Helmholtz's theorem (also exact on the Euler model) which states that *vortexlines move with the fluid*.

[†] It is, of course, because only $D\mathbf{u}/Dt$ (and not, for example, $\partial\mathbf{u}/\partial t$) is identified by eqn (71) as taking a simple form proportional to a gradient, that the theorem can be proved only for curves C moving with the fluid. Fortunately the one special complication which this brings into the proof is dealt with readily by use of eqn (141).

Such a statement has to be understood in the same sense as the statements about a closed curve C moving with the fluid in Section 5.1. It means, in fact, that the particles of fluid of which any vortexline is composed at any one instant move in such a way that the same chain† of particles of fluid continues to form a vortexline at all later instants.

To prove this theorem, we first demonstrate an equivalent result which is also useful in its own right: the property that *vortextubes move with the fluid* in the sense that the particles of fluid of which any vortextube is composed at any one instant move in such a way that the same tube-shaped assemblage of particles continues to form a vortextube at all later instants. Any vortextube, of course, is entirely made up of vortexlines (Fig. 23) so that this property would follow immediately if Helmholtz's theorem had been established. Conversely, if we knew that all vortextubes move with the fluid, we could deduce that each vortexline, being the limit of a very thin vortextube as its cross-section shrinks to zero, must share the same property.

Helmholtz's theorem is equivalent, then, to the result that all vortextubes move with the fluid. The truth of this result follows, however, from the fact that a vortextube is a tube-shaped surface with the vorticity everywhere tangential to it, i.e. such that

$$(\text{curl } \mathbf{u}) \cdot \mathbf{n} = 0, \qquad (148)$$

if $\mathbf{n}$ is a unit vector normal to that tube-shaped surface. This means, by Stokes's theorem (eqn (134)), that the circulation is zero around *any* closed curve C which lies on the surface of the tube *without embracing it* (Fig. 28); for such a closed curve is the boundary ∂S of a piece S of that tube-shaped surface on which eqn (148) is satisfied. The particles of fluid making up the vortextube form, therefore, at any *subsequent* time a tube-shaped assemblage of particles with the same property, i.e. that there is zero circulation around any closed curve C which lies on its surface without embracing it (since Kelvin's theorem equates that circulation around C to the circulation around the corresponding closed curve C which is made up of the same particles of fluid on the original vortextube). Finally, for the tube-shaped assemblage at that later time, we demonstrate at every point the characteristic property (eqn (148)) which shows it to be a vortextube, by taking C to be a very small loop around that point within which $(\text{curl } \mathbf{u}) \cdot \mathbf{n}$ is effectively constant so that eqn (148) follows from Stokes's theorem (eqn (134)).

The two mutually equivalent results (on vortexlines and vortextubes) which have just been demonstrated have dynamical implications for both the direction and the magnitude of the vorticity. Its *direction* turns as does that of the vortexline. More precisely, its direction turns as does the direction of the

---

† Here the word 'necklace' might have been used again as in Section 5.1 but for the fact that a vortexline need not be a *closed* curve.

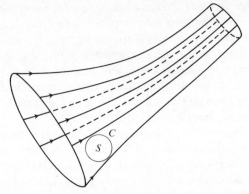

Fig. 28. Any closed curve C which lies on the surface of a vortextube without embracing it must be the boundary of a surface S on which eqn (148) is satisfied. The circulation around C is therefore zero. The same tube-shaped assemblage of particles therefore exhibits the same property for any such C at any later time and, by taking C to be a very small loop, we deduce that the assemblage is still necessarily a vortextube.

chain of particles envisaged in Helmholtz's theorem as composing the vortexline. At the same time, its *magnitude* increases or decreases in proportion to the stretching (elongation) or foreshortening of that chain of particles.

This last very important result follows from Kelvin's theorem applied to a thin vortextube of very small cross-sectional area S surrounding the vortexline. The circulation around any closed curve embracing this vortextube (defined in Section 4.5 as its 'strength', and equal for the thin vortextube to S times the magnitude of the vorticity) remains constant since that curve, as part of the vortextube, moves with the fluid. If, at the same time, there is any change (elongation or foreshortening) in the length l of a small cylindrical section of that vortextube, its cross-section S is bound to change (for a fluid of constant density) in such a way that the volume it occupies (S times the length l) remains constant. Finally, then, the magnitude of the vorticity changes in direct proportion to l (the only possibility if the product of each with S is to remain constant).

Stretching a vortexline makes the fluid spin faster, then, in the sense that the angular velocity $\frac{1}{2}\omega$ of the rigid-rotation constituent (Fig. 20) of a spherical particle's motion increases. We can appreciate why this is so from the standpoint of the general laws of mechanics: any pure *pressure* forces which, being necessarily directed normal to the surface of a spherical particle, act through its mass centre must have zero *moment* about the mass centre. This implies a zero rate of change of angular momentum about the mass centre as the sphere is subjected to the elongations and foreshortenings along principal axes of rate of strain illustrated in Fig. 20. The *component* of angular velocity about an axis that is being elongated therefore increases (so as to keep the corresponding angular momentum component constant) because the moment of inertia about that stretched axis is decreasing. Simultaneously, the moment of inertia about a foreshortened axis is increasing so that the component of angular velocity about such an axis is reduced. It can be demonstrated that these considerations, if carefully made quantitative, lead to the same conclusions on the changing magnitude and direction of the vorticity vector as those deduced above from Kelvin's theorem.

Besides being exact on the Euler model, Helmholtz's theorem shares with Kelvin's theorem a feature noted in Section 5.1, that any vortexline 'moves with the fluid' to a very close approximation provided that viscous forces *at the position of the vortexline itself* are very small. This is made evident either: (i) from the proof of Helmholtz's theorem by applying Kelvin's theorem to a small closed curve on a very thin vortextube around the vortexline; or (ii) from the angular-momentum argument applied to a small spherical particle (since only a viscous force at its surface could exert a moment about its mass centre). We shall, indeed, find (Section 5.5) that the type of modification to a vortexline's exact property of moving with the fluid, produced in a fluid of viscosity μ, is of the nature of a slow *diffusion* of each component of vorticity *down its gradient* with diffusivity μ/ρ.

In a 'smoke ring', the smoke particles mark a ring-shaped bundle of closely packed vortexlines, about which they make the fluid rotation readily visible. Stretching those vortexlines (e.g. by blowing the ring at the sharp point of a cone, up which it moves while necessarily being stretched) is observed to enhance the rotation. Finally, both the rotation and the smoke concentration are significantly *diffused* (the gradients of both being substantial) as time goes on.

5.3 Irrotational flow

Before pursuing the topic of vortexlines any further, however, we first devote some attention to a rather special, but nevertheless important, category of flows: those throughout which there is zero vorticity. These are flows such that, for any spherical particle of fluid, one of the three components of its motion illustrated in Fig. 20 is completely absent (the rotation). For this reason, such flows are called *irrotational*. In an irrotational motion, a small spherical particle of fluid is subjected only to a pure translation at the velocity of its mass centre and to a pure rate of strain.

The following theorem, foreshadowed by Lagrange and rigorously stated and proved by Cauchy, is usually called Lagrange's theorem. This important theorem, exact on the Euler model, as is Kelvin's theorem (from which we deduce it), states that *a flow irrotational at any one initial time is irrotational also at all later times.*

The theorem is proved by selecting any later time t and any point P within the fluid and any component of vorticity (say, the x component ξ) and showing that $\xi = 0$ at P at time t. For this purpose, consider a very small circular disc S with centre P and with its unit normal $\mathbf{n}$ in the x direction. Then in Stokes's theorem (eqn (134)) we have

$$(\text{curl } \mathbf{u}) \cdot \mathbf{n} = \xi; \tag{149}$$

and ξ varies insignificantly over the very small disc S from its value ξ_P at P. The circulation (in the positive sense relative to $\mathbf{n}$) around the closed curve ∂S forming the boundary of S is therefore

$$\int_{\partial S} \mathbf{u} \cdot d\mathbf{x} = \xi_P S. \tag{150}$$

But, by Kelvin's theorem, that circulation $\xi_P S$ around ∂S at time t is the same as the circulation around a closed curve C which consisted of the same particles of fluid at the *initial* time, when the vorticity was everywhere zero (that is, when the flow was irrotational). At that time, application of Stokes's theorem to the closed curve C shows the circulation $\xi_P S$ around C to be zero, so that $\xi_P = 0$; and, similarly, all components of vorticity can be proved to be zero at all points at all later times (Lagrange's theorem).

One important special case of Lagrange's theorem is concerned with the development of motions in which the fluid is initially at rest (so that $\mathbf{u} = 0$ everywhere). Then curl $\mathbf{u} = 0$ initially and, therefore, Lagrange's theorem tells us that, on the Euler model, *such motions started from rest are irrotational* at all later times.

There is great interest in many motions of fluid which start from rest (for example motions of initially still water generated by a fish beginning to swim, or by a ship beginning to move, or by swell from a distant storm travelling into such undisturbed water), so that the potential importance of the above conclusion is clear. The fundamental equation

$$\text{curl } \mathbf{u} = 0 \tag{151}$$

of irrotational motion is a simple, linear equation, which is much better suited (as we shall see) to the convenient computation of solutions than the complicated, nonlinear eqn (68).

We are, accordingly, most interested in analysing irrotational motions, as well as in studying those conditions when, in spite of viscous forces, motions started from rest are able to remain irrotational except within thin boundary layers (see Sections 3.4, 4.3, and 5.5) and wakes. The further discussion of the general properties of irrotational motions is, nevertheless, postponed to Chapter 6; being preceded by some further analysis of the properties, and the dynamics, of vorticity which can be of the greatest help towards identifying which motions of fluids can be expected, in the main areas of the flow, to be irrotational.

5.4 Line vortices and vortex sheets

The chief practical restriction on the permanence of irrotational flows (demonstrated in Lagrange's theorem to be exact on the Euler model) is associated with the conditions at a solid boundary which have been discussed in Section 3.4. In that Section, the single, self-evident boundary condition (eqn (74)) on the normal component of velocity was stated to be sufficient, on the Euler model, to determine a flow fully. Indeed, for initially irrotational motions (such as motions started from rest), this condition (eqn (74)) is proved in Section 6.2 (on the Euler model) to specify uniquely the subsequent irrotational motion at each instant; an irrotational motion which, as already mentioned, can conveniently be computed.

On the other hand, the true condition at a solid boundary (Section 3.4) equates not only the normal components of the velocities of the fluid and of the solid surface as in eqn (74) but also their tangential components. The effect of this is to destroy any exact permanence of irrotational flows, essentially (as we shall see) through the *generation of vorticity* at any such solid surface. This section, and the next two, are concerned with indicating how such new vorticity is created at a solid surface, and with discussing how the associated vortexlines may be convected with the fluid (in accordance with Helmholtz's theorem) and also diffused (by the action of viscosity).

Often, the convection and diffusion of vortexlines which initially emerged from the solid boundary are found to leave them concentrated in a very thin region. It may, in many cases, be thin in just one of its dimensions, as a sheet is, and then it is often called a *vortex sheet*. A boundary layer (see Sections 3.4 and 5.5) is an example of a vortex sheet. Alternatvely, some vortexlines that have been convected away from the solid boundary may, in certain circumstances, become concentrated in a region thin in *two* of its dimensions (essentially just in the immediate neighbourhood of a single line), and such a concentration of vorticity around a single line is usually called a *line vortex*.

The properties of line vortices and vortex sheets, now to be investigated, are important because they are simple. This may, often, allow us to make ready calculations of flows in which all of the vorticity is concentrated in very thin regions. Such calculations use the simple eqn (151) for irrotational motion outside those regions, alongside additional conditions derived from the properties studied below.

A line vortex is a concentration of vorticity within a small distance, say ε, of a particular (straight or curved) line. We now consider its relationship to an external flow which is irrotational. This means that *all* of the motion's vortexlines are concentrated within the tube of radius ε with the line as axis; or, if we prefer to think in terms of vortextubes, that tube is a bundle (straight or twisted) of vortextubes which are stretched along it. Evidently, a line vortex 'moves with the fluid' since (Section 5.2) both vortexlines and vortextubes have this property.

We now define the 'strength' of a line vortex, by analogy with the definition (Section 4.5) of the strength of a vortextube, and, indeed, in such a way that the line vortex's strength is just the sum of the strengths of all the vortextubes in the bundle of which we may think of it as made up. Quite simply, the strength of a line vortex is defined as the circulation around any closed curve C which lies in the irrotational flow outside it and which embraces the line vortex exactly once in a particular sense. Usually, the sense in question is chosen so that the circulation around C in that sense is a *positive* number $+K$ (whereas the circulation in the opposite sense would be $-K$).

Figure 29 demonstrates that this is an unambiguous definition by taking any two such closed curves C_1 and C_2 and proving that the circulations K_1 and K_2 around them are equal. This is achieved by applying Stokes's theorem (eqn (134)) to a surface S entirely within the irrotational flow (so that the left-hand side of eqn (134) in zero) and in the shape of a collar (a finite length of tube) with its boundary ∂S made up of C_1 and C_2. Whichever direction is

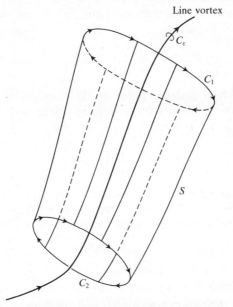

Fig. 29. The circulations K_1 and K_2 around two different closed curves C_1 and C_2, that each embrace a line vortex exactly once in the *same* sense, are necessarily equal. This is proven by applying Stokes's theorem to a collar-shaped surface S whose boundary ∂S consists of C_1 and C_2 taken in *opposite* senses.

chosen for the unit normal $\mathbf{n}$ to S, the circulations around C_1 and C_2 taken in the positive sense relative to $\mathbf{n}$ are *either* $+ K_1$ and $- K_2$ *or* $- K_1$ and $+ K_2$, as Fig. 29 indicates; therefore, putting the right-hand side of eqn (134) equal to zero shows that $K_1 = K_2$.

Any choice of C, then, gives the same value K for the line vortex's strength. In particular, taking C as a simple circle C_ε of radius ε which (Fig. 29) is the boundary of a disc-shaped cross-section S_ε of the line vortex shows, by eqn (138), that

$$K = \int_{S_\varepsilon} \boldsymbol{\omega} \cdot \mathbf{n} \, dS, \qquad (152)$$

where the unit vector $\mathbf{n}$, being normal to the cross-section S_ε, is tangential to the line vortex itself. Equation (152) tells us that the strength K of any line vortex is an integral of the tangential component of vorticity over its cross-section; while, if the line vortex is thought of as a bundle of vortextubes all threading through the cross-section S_ε, it tells us that the sum of their strengths (Section 4.5) is K.

At the same time, Kelvin's theorem (eqn (147)) identifies the strength K as an unchanging feature of a line vortex. By contrast, its effective radius ε may change with time, increasing as a result of diffusion but decreasing as a result of stretching of the vortexline (Section 5.2).

The fluid velocity rises steeply near any line vortex. We can see this by writing K as the circulation

$$K = \int_{C_r} \mathbf{u} \cdot d\mathbf{x} \qquad (153)$$

around C_r, which is another curve in the form of a circle in a plane perpendicular to the line vortex, but this time with radius $r > \varepsilon$. By expression (137), we may write K as the integral

$$K = \int_{C_r} (\mathbf{u} \cdot \mathbf{t}) \, ds \qquad (154)$$

of the tangential component of fluid velocity with respect to arc length s along the circumference of C_r. That circumference has magnitude $2\pi r$, so that eqn (154) identifies the *average* tangential component $\mathbf{u} \cdot \mathbf{t}$ of fluid velocity as

$$\mathbf{u} \cdot \mathbf{t} = K/(2\pi r) \qquad (155)$$

for all $r > \varepsilon$. Figure 30 plots this average velocity tangential to such a circle of radius r as a function of r, and shows how eqn (155) represents a steep rise as r decreases within the region $r > \varepsilon$ of irrotational flow.[†]

[†] Typical motions outside line vortices are combinations of that depicted in Fig. 30 with other (usually, far more gradually varying) irrotational flows.

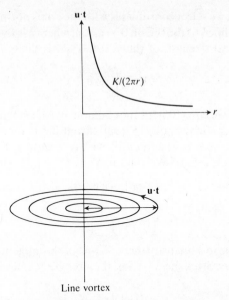

Line vortex

Fig. 30. A plot of the average tangential velocity component $\mathbf{u} \cdot \mathbf{t}$ near a line vortex as a function of the distance r from it, for values of r exceeding the maximum radius ε within which vorticity is significantly present.

It may at first be thought surprising that a motion like that depicted in Fig. 30 can be *irrotational* in the region $r > \varepsilon$ even though particles of fluid are orbiting around the line vortex. There is, however, no occasion for surprise. The condition for a motion to be irrotational requires only that each and every spherical particle of fluid has zero angular momentum, and the velocity distribution (eqn (155)), which increases towards the axis, achieves this. Indeed, the high angular momentum associated with *rigid* rotation at angular velocity Ω requires a tangential velocity

$$\mathbf{u} \cdot \mathbf{t} = \Omega r, \tag{156}$$

which *decreases* towards the centre. That angular momentum is completely cancelled out by the opposite angular momentum associated, as shown in Fig. 21, with the difference in rate of shear between eqns (155) and (156).

Although the line vortex is a valuable concept, an even more important concept is that of a *vortex sheet* in which the vortexlines are confined, not within a very thin tube, but within a very thin sheet. Just as the integral (eqn (152)) of vorticity over a cross-section defines the strength K of a line vortex, so the local strength of a vortex sheet is defined as an integral of the vorticity across the sheet's *thickness*.

The relevance of this strength can be seen most readily if, at a particular point of the sheet, we select coordinates with the z axis normal to the sheet and with vorticity confined to the region $0 < z < \delta$, where δ is the sheet's very small thickness. The local strength of the vortex sheet is then

$$\mathbf{V} = \int_0^\delta \boldsymbol{\omega} \, \mathrm{d}z. \tag{157}$$

But the expression in eqn (105) for the components of $\boldsymbol{\omega} = \operatorname{curl} \mathbf{u}$ contains only two terms which can make eqn (157) significant if δ is very small, namely the differential coefficients $\partial v/\partial z$ and $\partial u/\partial z$ with respect to z. Their contributions to the strength (eqn (157)) take the form

$$(-[v], +[u], 0), \tag{158}$$

where the square bracket

$$[\ \ ] = [\ \ ]_{z=0}^{z=\delta} \tag{159}$$

signifies the change in a quantity (here, in one of the tangential components of velocity) across the vortex sheet.[†] Thus, the strength $\mathbf{V}$ of a vortex sheet is equal to the change

$$([u], [v], 0) \tag{160}$$

in tangential velocity across the sheet, *turned through* 90° about the z direction so as to give expression (158).

We see, then, that the 'slipping' which occurs at a vortex sheet (a change in tangential velocity across the sheet) is in the direction which would be permitted by rigid rollers rotating about an axis in the direction of the integrated vorticity (eqn (157)), i.e. in a direction at right angles to that axis. Figure 21 reminds us, however, how different the real shearing motion of a fluid is from any simple rigid rotation: the rigid rotation is present, with vorticity equal to the rate of shear s (so that the integrated vorticity is equal, as above, to the overall velocity change), but that rotation is always accompanied by a pure rate of strain which enormously modifies its kinematic effect.

Often, in theoretical fluid mechanics, a vortex sheet of given strength and given thickness is modelled, for the sake of simplicity, as a vortex sheet of the same strength and *zero* thickness, i.e. as a *surface* across which the tangential velocity is discontinuous (a 'surface of slip'). Similarly, a line vortex of given strength and thickness may be modelled as a line vortex with identical strength K but with zero thickness, i.e. as a line where the velocity tangential to a small circle of radius r around it behaves asymptotically like eqn (155) as $r \to 0$. These can be useful idealizations provided that we do not forget in either case the nonzero thickness of the real structure, which must be present if only because of viscous diffusion.

[†] By contrast, the integral of one of the x or y derivatives in eqn 105) could be of comparable magnitude to $[u]$ or $[v]$ only if those derivatives were of order $[u]/\delta$ or $[v]/\delta$, which for very small δ would imply enormous changes of velocity after substantial changes in x or y.

5.5 Boundary layers

Diffusion, which plays a major role in determining the detailed structure of boundary layers, must now be described in general terms. A physical quantity or its *density* (value of the quantity per unit volume) is stated to be subject to diffusion when some physical effect is producing a rate of transfer of that quantity down its gradient, the rate of transfer of the quantity across unit area being equal in magnitude to the gradient of the quantity's density, multiplied by a coefficient called the *diffusivity*.

A good example is provided by conduction of heat in a homogeneous solid. Heat is transferred down the gradient of T (the temperature), with the heat transfer rate across unit area being equal to the conductivity k times the magnitude of grad T. But the gradient of the *heat per unit volume* (that is, the 'density' of the physical quantity, heat) is ρc grad T, where c is the specific heat (heat addition required to give unit *mass* a unit increase in temperature) and ρ is the mass per unit volume. Therefore the diffusivity of heat is the ratio

$$\kappa = k/\rho c, \tag{161}$$

that is the coefficient which when multiplied by the gradient of the heat per unit volume gives the magnitude of the heat transfer rate per unit area.

The definition of diffusivity (quantity transferred per unit area per unit time, divided by a gradient of the same quantity per unit volume) necessarily gives it the dimensions

$$(\text{length})^2/(\text{time}). \tag{162}$$

For the diffusivity of heat κ, convenient units are $\text{mm}^2\,\text{s}^{-1}$ and values for different solids range from a maximum of 171 for silver, through about 12 for iron, to about 0.6 for glass and still less for good thermal insulators.

Diffusion has the property that it smooths out any spatial discontinuity, converting it into a continuous transition region whose thickness increases with time. The diffusivity's dimensions (expression (162)) suggest that in time t after such a discontinuity is formed (perhaps, by a cold body being immersed in boiling water) the thickness of the transition region (the warmed part of the body) should initially grow in proportion to

$$(\kappa t)^{\frac{1}{2}}, \tag{163}$$

where κ is the diffusivity. This expectation is borne out in practice (Fig. 31) for any quantity subject to the effects of diffusion by themselves.

In fluid mechanics, the physical effect of diffusion competes with other physical effects (such as convection) and this produces more complicated behaviour. Nevertheless, some analogous indications from dimensional considerations are still very helpful.

The definition of the viscosity μ (Section 3.4) shows it to produce diffusion

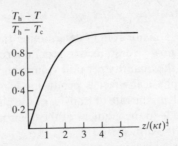

Fig. 31. Diffusion of heat, and of the temperature T, through a distance z into a cold body initially at temperature T_c after a time t has elapsed since it was suddenly immersed in hot liquid at temperature T_h. Heat has diffused through a distance proportional to $(\kappa t)^{\frac{1}{2}}$ where κ is the diffusivity.

of *momentum*, with a diffusivity (also sometimes called the kinematic viscosity)

$$v = \mu/\rho. \tag{164}$$

This is because the fluid momentum per unit volume is ρ times the velocity, while the viscous force per unit area (which, since force is rate of change of momentum, represents a rate of transfer of momentum across unit area) is equal to μ times the gradient of velocity. Diffusion of momentum (or, equally, of the velocity **u**, which is proportional to the density of momentum) is therefore present in addition to the physical effects taken into account in the Euler model.[†]

Now the definition (expression (104)) of the vorticity $\boldsymbol{\omega} = \text{curl } \mathbf{u}$ allows us to infer that it, too, must be subject to diffusion (with diffusivity $v = \mu/\rho$), in addition to physical effects allowed for in the Euler model. This is because each term in the expression (104) for $\boldsymbol{\omega}$ is a gradient of one of the velocity components, and therefore its change due to viscosity must correspond to diffusion (with this diffusivity), simply because this is so for the velocity components themselves. Since the effects on the vorticity field occurring on the Euler model are those described by Helmholtz's theorem (that vortexlines move with the fluid), we finally infer that, *while* vortexlines move with the fluid (are convected with the fluid), they are simultaneously subjected to diffusion with diffusivity v.

The value of the diffusivity $v = \mu/\rho$ at atmospheric pressure is greater for air than for water, as the following table shows. Broadly speaking, this is because, although the viscosity μ is greater for water than for air by about two orders of magnitude, the density ρ, which controls the dynamic effect of viscous forces, is greater for water than for air by about *three* orders of magnitude.

[†] Mathematically, this is represented by the extra term $\mu\nabla^2\mathbf{u}$ mentioned in the footnote on p. 37 as appearing on the right-hand side of the Euler equation (eqn (68)).

Diffusivity v in mm^2 s^{-1} at atmospheric pressure

	Temperature (°C)				
	0	10	20	30	40
Air	13.2	14.1	15.0	15.9	16.8
Water	1.79	1.30	1.00	0.80	0.66

As in expression (163), dimensional considerations suggest that a vortex sheet which initially has zero thickness, and so represents a 'surface of slip' (see end of Section 5.4) or discontinuity in tangential velocity, may be expected to develop into a continuous layer whose thickness at first grows in proportion to

$$(vt)^{\frac{1}{2}}. \tag{165}$$

We consider now how well this expectation corresponds with observations of: (i) vortex sheets close to solid boundaries; and (ii) free vortex sheets (Section 5.6).

The first of these two cases is particularly important because of the boundary condition at a solid boundary (Section 3.4). Suppose (say) that in initially undisturbed fluid a solid body suddenly starts to move. According to the Euler model, this at once sets up an irrotational flow, involving a nonzero tangential motion of fluid (relative to the motion of the body) at its solid boundary. There is, accordingly an initial discontinuity between the tangential velocity of fluid in that external irrotational flow and the actual zero value of the tangential velocity of fluid (relative to the motion of the body) in direct contact with the boundary. Such a discontinuity can be described as a vortex sheet, initially of zero thickness, attached to the solid boundary. Subsequently, the sheet's thickness may be expected to grow by diffusion, initially in proportion to expression (165).

Figure 32 illustrates such a *boundary layer* in a frame of reference in which the body is at rest (so that all velocities indicated in Fig. 32 are velocities relative to the body). Note that the effect of diffusion which causes the layer thickness to increase in proportion to $(vt)^{\frac{1}{2}}$ is combined, as noted earlier, with the effect of convection (that is, of vortexlines moving with the fluid). It is the counterplay of these two effects that gives boundary layers several important and characteristic properties which are described in the rest of this chapter.

Fig. 32. A boundary layer in flow around a body at rest. Vorticity has diffused through a distance proportional to $(vl/U)^{\frac{1}{2}}$ where v is the diffusivity and where the time for fluid to flow around the body is proportional to l/U.

First of all, if U is the velocity of the body relative to the undisturbed fluid, then U also represents a typical magnitude of the velocity of the fluid at the outer edge of the boundary layer in the frame of reference of Fig. 32 (that is, relative to the body). The vortexlines in the outer part of the boundary layer are being carried along by the fluid, then, at a velocity whose magnitude is around U; and, for a body of length l, they are swept clear of the body after a time t which is around l/U. This tends to limit the growth in boundary-layer thickness to a value proportional to

$$(vl/U)^{\frac{1}{2}}, \tag{166}$$

essentially because vorticity generated at the solid surface is swept away after a time around l/U, when by expression (165) it has diffused to such a distance. That vorticity must, of course, be replaced by new vorticity generated at the solid surface, so that the overall strength of the vortex sheet maintains the magnitude of 'slip' needed by the external irrotational flow, and this new vorticity once more diffuses a distance proportional to expression (166) before being swept away.

The importance of flows in which boundary layers can be very thin in relation to a dimension (say l) of the body to which they are attached was stressed in Section 3.4. Expression (166) now tells us that the ratio of l to the boundary layer thickness can be very large only if the quantity

$$(Ul/v)^{\frac{1}{2}} \tag{167}$$

is very large. This is the square root of the *Reynolds number*,

$$R = Ul/v, \tag{168}$$

a non-dimensional measure of the flow speed and of the scale of the body in relation to the magnitude of viscous diffusion effects. Evidently, the square root $R^{\frac{1}{2}}$ can be 'very large' only if R itself is very large indeed; but values

$$R > 10^4 \tag{169}$$

are large enough for boundary layers to remain very thin. This criterion influenced the comment in the Preface which restricted this book's main subject matter to flow speeds greater than about 1 m/s (10^3 mm/s) which, with size l of body around 10^2 mm or more, would tend to satisfy the inequality (169) for the values of v quoted above.

It was Osborne Reynolds (1842–1912) who first introduced the important concept of the Reynolds number (eqn (168)). His experiments from 1883 onwards showed the significance of this number, *not* for the boundary layers which were to be discovered rather later by Ludwig Prandtl (1875–1953), but in relation to the chaotic form of motion known as turbulence. Reynolds demonstrated for a variety of particular types of motion of fluids that the onset of chaotic or turbulent motion occurred when the Reynolds number exceeded a particular critical value.

Much later, experiments showed furthermore that, although irrotational motions are immune to turbulence, any boundary layer may become turbulent at a *sufficiently* large value of the Reynolds number. The required value depends on various factors (such as boundary shape) but is typically rather nearer to 10^6 than to the limit of 10^4 noted in the inequality (169). The random motions of turbulence increase rates of transfer of quantities down their gradients and thus enhance the effective diffusivity above the 'viscous' value $v = \mu/\rho$. But although this increases the boundary layer thickness to a value significantly above that in expression (166), nevertheless the boundary layer at these Reynolds numbers continues to be very thin compared with l.

5.6 Separation

We have seen in Section 5.5 that, because diffusion is a slow process, its direct effect upon the growth of boundary layers on solid boundaries in flows of fluid at speeds substantial enough to satisfy the condition in the inequality (169) may be extremely limited, thus allowing them to remain very thin. If that happens, the flow outside such a boundary layer may be predicted quite well with the relatively simple Euler model.

On the other hand, we have emphasised many times (see especially Sections 1.3 and 3.4) that in a wide class of flows no such conclusion can be drawn because the flow *separates* from the solid surface. It is, in fact, the counterplay of the convection and diffusion of vorticity within the boundary layer which can cause it to become separated from the solid boundary in certain circumstances, which we must now try to analyse.

Figure 33 shows a steady flow (Section 3.5) around a solid body, the thickest section of which significantly compresses the streamtubes in the motion outside the boundary layer. The velocity within any one of those streamtubes grows, therefore (Section 2.4), to a maximum at the place where the streamtubes are most compressed, and is reduced again beyond that place. This means that, if the boundary layer is to be considered as a vortex sheet attached to the solid surface, the local strength (see eqn (157)) of that vortex sheet has a magnitude V (the total slip permitted by the boundary layer) which increases up to the body's thickest point and decreases beyond that point. Note that V can be viewed both as the integral of vorticity across the thickness of the boundary layer and as the velocity of flow just outside the boundary layer.

For two points A and B within the region of accelerating flow in Fig. 33, the strength V is greater at the downstream point B than at the upstream point A. In the meantime, convection (the property that vortexlines move with the fluid) is removing vorticity from B at a greater rate than vorticity from A is replacing it, both because the integrated vorticity V is *greater* at B than at A and because the velocity of convection of that vorticity varies in the layer between zero at the solid surface and a maximum value V which is higher at B than at A. Yet the actual strength of the vortex sheet at B, which this convection

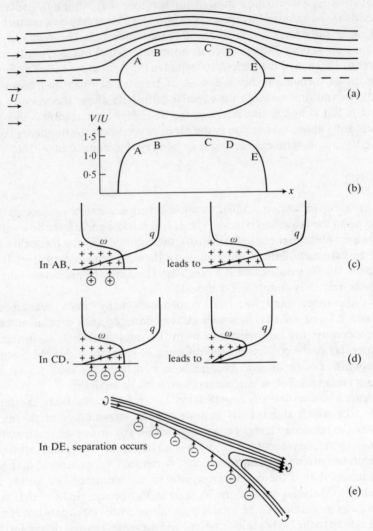

Fig. 33. Separation of a boundary layer for flow around an elliptic cylinder at rest in an oncoming stream of velocity U directed along the major axis of its elliptic cross-section.

(a) Streamlines, calculated by a method described in Chapter 9, with the flow assumed irrotational in the region outside the boundary layer. The broken line is the plane of symmetry of the flow.

(b) Distribution of the velocity V just outside the boundary layer, as given by the same irrotational-flow calculation. The boundary layer, then, consists of a vortex sheet of strength V.

(c) Between A and B, this strength V is increasing, thus requiring the generation of new positive vorticity at the boundary so that the distributions of vorticity ω and fluid speed q within the boundary layer are changed as shown (with the boundary-layer scale normal to the body surface hugely exaggerated).

effect is tending to reduce, must remain equal to the external flow velocity at B so that this external flow is permitted to slip over the surface. Accordingly, new vorticity is necessarily being generated at the solid surface with the *same* sense of rotation as the vorticity in the boundary layer at B, in order that the necessary value of the integrated vorticity V at B can be maintained. This new vorticity is in turn convected and diffused, but the whole process just described is fully consistent with the arguments used in Section 5.5 to estimate boundary layer thickness as a quantity proportional to expression (166).

When, however, we apply the same arguments to two points C and D in the region of Fig. 33 where the flow is being *retarded*, so that V is smaller at D than at C, we obtain a different conclusion. Convection is removing vorticity from D at a rate smaller than that at which vorticity from C is replacing it (since V, which is the integrated vorticity *and* the maximum convection velocity, is smaller at D than at C), and yet (as before) the actual strength of the vortex cannot change. Then at a point, such as D, where convective effects are adding vorticity, there is necessarily new vorticity being generated at the solid surface with the *opposite* sense of rotation to that in the boundary layer.

On the other hand, any relatively modest rate of generation of vorticity in the opposite sense such as may occur at a point D where the flow is only *slightly* retarded can be more than cancelled out by the rate of diffusion of the main boundary-layer vorticity down its gradient towards the solid surface. In such a case, there is no actual appearance of any vorticity with the opposite sense of rotation at D.

The situation is quite different, however, at a point E where the flow is being *strongly* retarded. There, convection is removing vorticity at a rate very much smaller than the rate at which vorticity from D is replacing it and, in consequence, new vorticity with the opposite sense of rotation must be generated at the solid surface at a rate greater than can be compensated for by vorticity diffusion from within the boundary layer. At E, then, a *reversed flow* develops near the boundary (that is a shearing flow with opposite rate of shear, associated with an opposite sense of the vorticity). At a point on the solid surface somewhere between D and E the forward flow at D and the reversed flow at E must meet, and become separated from the surface. The main external flow continues past the body but with a thick region of reversed flow (rather than a very thin boundary layer) lying between it and the solid surface.

(d) Between C and D, the same vortex-sheet strength V is slowly decreasing, requiring the generation of new negative vorticity at the boundary, but only in modest amounts which are counteracted by diffusion of positive vorticity towards the surface from the main part of the boundary layer.

(e) Between D and E, however, a much bigger generation of negative vorticity at the boundary overcomes the effect of diffusion. Vorticity near the boundary then acquires opposite sign, corresponding to reversed flow. Where this meets the oncoming flow, both become separated from the surface.

In the above description, the main ideas of vortex dynamics have been used to give as simply as possible an interpretation of the data on when and where flow separation is observed to occur. In summary, boundary layers show no tendency to separate where there is an accelerating external flow. Where the external flow is weakly retarded, the tendency to separation associated with a modest rate of generation of vorticity of opposite sign at the surface is overcome by the rate of diffusion of the main boundary layer vorticity towards that surface. On the other hand, where the external flow is *strongly retarded*, the flow separates because the rate of diffusion cannot overcome the then much greater rate of generation of vorticity in the sense associated with reversed flow.

At those substantially greater Reynolds numbers for which a boundary layer is turbulent (Section 5.5), the rate of diffusion is increased. Under these circumstances a still stronger retardation of the external flow is needed before separation occurs. Thus, if a boundary layer becomes turbulent, its separation is postponed. It still separates, however, when the external flow is retarded sufficiently strongly.

This is the background against which conclusions from the Euler model are evaluated in subsequent chapters. Where those conclusions predict that a flow along a solid surface is accelerated, or weakly retarded, they can be expected to represent well the flow outside a boundary layer which, on the assumption of the inequality (169), should be very thin and remain attached to the surface. Where, however, they predict that flow along a solid surface is strongly retarded, the real flow must be expected to show major differences from those predicted on the Euler model because the flow separates from the solid surface. A relatively stronger retardation is necessary to cause separation, on the other hand, if the boundary layer has become turbulent.

This chapter may now be concluded with a brief note on a topic not much pursued in subsequent chapters: the general nature of flows that have undergone separation. Immediately downstream of separation, we often observe something like a *free* vortex sheet, where the separated external flow slips past a region of very much slower recirculating motion including the reversed flow already referred to (Fig. 33). However, all free vortex sheets are found in practice to be extremely unstable: even at very moderate Reynolds numbers they readily break up into discrete line vortices (Fig. 34); further-

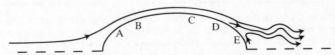

Fig. 34. The instability of the free vortex sheet produced by separation of the flow around the upper half of the elliptic cylinder shown in Fig. 33. Note: although the broken line represents the plane of symmetry for the irrotational flow of Fig. 33(a), nevertheless the onset of instability makes the flow cease to be accurately symmetrical about this broken line.

more, they tend to develop into turbulence involving chaotic motions of a particularly vigorous nature when the Reynolds numbers surpass limits that are much less than for boundary layers. A lot of energy is dissipated in such motions, which is why engineering design prefers in general to aim at achieving the flows that do not separate and which are mainly studied in the chapters that follow.

6
General properties of irrotational flows

In Chapter 5, irrotational flows were defined as flows throughout which the vorticity is zero (Section 5.3), while those conditions (especially, boundary conditions) which uniquely determine an irrotational flow were briefly indicated (at the beginning of Section 5.4). The present chapter is concerned with firmly establishing sets of conditions which, at a given moment in a given region of fluid, specify instantaneously a unique irrotational flow of the fluid. The chapter identifies also a particular characteristic (minimum kinetic energy) which distinguishes that flow from any flow with nonzero vorticity which also satisfies the same conditions; conditions which, in addition, are shown to specify the latter flow uniquely provided the vorticity field is given. In regions of a certain type, called *doubly connected*, the conditions for uniqueness are found to require supplementation with a certain *circulation* condition.

Irrotational flows are important because fluid motions which start from rest may remain very close to the irrotational flow predicted (Section 5.3) by the Euler model within systems designed such that boundary-layer separation (Section 5.6) is avoided. A particular property of irrotational flows, the existence of a *velocity potential*, greatly facilitates their calculation, while also allowing the Bernoulli relationship between the velocity field and the pressure field which is normally restricted to steady flows (Section 3.5) to be modified so as to become applicable in the general case of unsteady flow fields when these are irrotational.

6.1 Pressure and velocity potential

One type of flow which is necessarily irrotational is the instantaneous flow set up initially, in a region of fluid at rest, by the impulsive motion of a solid boundary. It is for this special type of irrotational flow that the existence of a velocity potential and its relationship to the pressure field will first be established.

An 'impulse' represents, of course, the effect of a large force applied for just a very short time, as when a solid body is given a knock which sets it moving. Quantitatively, the impulse represents the total momentum imparted. During the very short time in question (say, from $t = 0$ to $t = \tau$) the *rate* of change of that momentum is the force $\mathbf{F}$ applied. Accordingly, the impulse $\mathbf{I}$, or total change of momentum during time τ, is the integral,

$$\mathbf{I} = \int_0^\tau \mathbf{F} \, \mathrm{d}t, \tag{170}$$

of the large force $\mathbf{F}$ over the very short time τ of its application. Mathematical models make fruitful use of the idea of impulse in cases when the momentum change $\mathbf{I}$ is a substantial one, even though the time of application τ can (in the context of the motion as a whole) be regarded as negligibly small and can effectively be replaced by zero.

In a region of fluid initially at rest but set into motion by the impulsive movement of a solid boundary, all the vorticity present must be generated at the boundary and be subsequently diffused and convected (Section 5.5). In a negligibly small time τ, however, such diffusion and convection of vorticity can only move it a negligible distance away from that solid boundary, so that the 'boundary layer' (Section 5.5) remains negligibly thin. Essentially, then, the entire fluid motion instantaneously set up by the boundary's impulsive motion must *initially be irrotational*. This statement remains true, and potentially important, even though the boundary layer's later growth and (possible) separation may greatly alter the flow's character from the initially irrotational type.

Simple mechanics, using arguments like those leading to eqn (170) for change of momentum, gives a still more useful form of the same conclusion. If the velocity field instantaneously set up is $\mathbf{u}$, then the momentum imparted to each particle of fluid is $\rho\mathbf{u}$ per unit volume. The total *force* per unit volume acting on the particle may, however (see eqn (71)), be written $-\operatorname{grad}p_e$, where p_e is the excess of the pressure over its hydrostatic value. Accordingly, eqn (170) when applied to a particle of fluid gives the instantaneously imparted momentum as

$$\rho\mathbf{u} = -\int_0^\tau (\operatorname{grad} p_e)\,dt. \tag{171}$$

Equation (171) implies that the velocity field can be written as the gradient

$$\mathbf{u} = \operatorname{grad}\phi \tag{172}$$

of a *velocity potential*[†] ϕ; where, for this impulsively started flow,

$$\phi = -\rho^{-1}\int_0^\tau p_e\,dt. \tag{173}$$

The check that the initial motion (eqn (172)) is indeed irrotational (in other words, that $\operatorname{curl}\mathbf{u} = 0$) is then immediate from the general vector identify

$$\operatorname{curl}(\operatorname{grad}\phi) = 0; \tag{174}$$

[†] Historically, the term 'velocity potential' was introduced by analogy with the potential of a conservative field of force. Accordingly, in early texts, ϕ is defined through an equation like eqn (172) but with a minus sign (that is $\mathbf{u} = -\operatorname{grad}\phi$) just as eqn (33) specifies gravitational force as minus the gradient of its potential. Because, however, the analogy is a tenuous one and the minus sign is a little inconvenient in practical calculations, the velocity potential has for many decades been defined through the velocity field being equated to plus its gradient.

an identity which holds for any function ϕ with continuous second derivatives, because eqns (32) and (105) make, for example, the x component of curl $(\text{grad}\,\phi)$ equal to $\partial^2 \phi/\partial z \partial y - \partial^2 \phi/\partial y \partial z$ (which is zero for such a function).

Some of the above conclusions for impulsively started flow may at first sight seem surprising. Equation (172), for example, implies that ϕ is a quantity of substantial magnitude (whose gradient is the velocity field), yet eqn (173) makes it proportional to the integral of p_e over just a very short time τ. These facts are compatible only if, during that time, the excess pressure field p_e has very *large* values. It must be recognized, then, that just as the solid boundary is subjected to a total impulse which requires large forces to act on it for a very small time τ, so this impulse produces throughout the fluid comparably large excess pressures p_e for which the integral in eqn (173) has a substantial value, even when taken over just the very short time τ. This in turn is why, for impulsively started flow, the equation of motion (eqn (71)) on being integrated from $t = 0$ to $t = \tau$ yields the result in eqn (171) just as it would if the left-hand side were $\rho \partial \mathbf{u}/\partial t$. Quite simply, the very large excess pressure gradient $(-\text{grad}\, p_e)$ completely dominates over the moderately sized difference $\rho \mathbf{u} \cdot \nabla \mathbf{u}$ between $\rho D\mathbf{u}/Dt$ and $\rho \partial \mathbf{u}/\partial t$.

It may, again, be thought surprising that such large excess pressures p_e can be generated throughout the region of fluid in an *extremely short time* τ by impulsive motion of a solid boundary. Physically, the reason is that pressure changes are transmitted through a fluid at an extremely high speed: the speed of sound. Indeed, the neglect of compressibility in the Euler model is equivalent to regarding the speed of sound as infinite, so that these pressure changes are transmitted throughout the region of fluid without any delay. (Furthermore, the real time required for a pressure change to be transmitted over a distance of, say, 10 m is only 0.03 s for air and 0.007 s for water.)

The existence of a velocity potential ϕ that satisfies eqn (172), which by eqn (174) requires the velocity field $\mathbf{u}$ to be irrotational, is established above for impulsively started flows. Furthermore, for these flows, the relation between ϕ and the excess pressure p_e takes the form of eqn (173). For more general irrotational flows it will be shown (Section 6.2) that a velocity potential ϕ with $\mathbf{u} = \text{grad}\,\phi$ does necessarily exist if the region occupied by the fluid is of a certain type, which is called *simply connected*. For such general flows with a velocity potential ϕ (flows which may *not* have been started impulsively from rest), the relationship between ϕ and the excess pressure is, however, a little more complicated.

The additional complication is derived from the impossibility, for flows not started impulsively, of neglecting the second term in the equation of motion

$$\partial \mathbf{u}/\partial t + \mathbf{u} \cdot \nabla \mathbf{u} = -\rho^{-1} \,\text{grad}\, p_e \qquad (175)$$

derived from eqns (71) and (64). For general irrotational flows, however, the discussion from eqn (65) onwards demonstrates a remarkable fact, that the second term in eqn (175) can simply be written as $\nabla(\frac{1}{2}\mathbf{u}\cdot\mathbf{u})$. Thus, the components in expressions (66) and (67), written out in detail to show that these quantities are in general unequal, are seen to coincide exactly when the components (in eqn (105)) of curl $\mathbf{u}$ are all zero. In short, *irrotational* flows satisfy the equation

$$\mathbf{u}\cdot\nabla\mathbf{u} = \text{grad}\,(\tfrac{1}{2}q^2), \tag{176}$$

even though it is only possible for flows in general to state, as in eqn (83), that the scalar products of each side of eqn (176) with $\mathbf{u}$ are identical.

Flows for which a velocity potential ϕ exists, with $\mathbf{u} = \text{grad}\,\phi$, are necessarily irrotational; this allows eqn (176) to be used in eqn (175) which therefore becomes

$$\text{grad}\,(\partial\phi/\partial t + \tfrac{1}{2}q^2 + \rho^{-1}p_e) = 0. \tag{177}$$

The quantity in brackets in eqn (177) must be independent of x, y and z, since its gradient is everywhere zero. It can at most be a function of t alone, therefore, yielding the relationship between velocity potential and excess pressure in the classical form

$$\partial\phi/\partial t + \tfrac{1}{2}q^2 + \rho^{-1}p_e = f(t). \tag{178}$$

The exceedingly useful eqn (178) governing unsteady irrotational flows takes simplified forms in two extreme cases. One of these is *steady* irrotational flow, in which the pressure and velocity fields are independent of time (Section 3.5) so that the velocity potential, too, can be taken independent of the time t. The first term in eqn (178) then disappears, and the second and third terms are independent of time so that the right-hand side must simply be a constant throughout the flow:

$$\tfrac{1}{2}q^2 + \rho^{-1}p_e = \text{constant throughout the flow.} \tag{179}$$

This is a completely new result, although its form may seem familiar from Section 3.5. There, for steady flows in general, the Bernoulli equation (eqn (86)) identified $p_e + \tfrac{1}{2}\rho q^2$ as taking a constant value along any streamline. For *irrotational* steady flows, however, we have shown that the *same* constant value of this expression is taken on *all* streamlines. This is a most important extension of the Bernoulli equation; an extension which, itself, is further extended to the case of *unsteady* irrotational flows in eqn (178).

Another extreme case of eqn (178) identifies the very large excess pressures momentarily present during the instantaneous start up of a flow through impulsive motion of the boundary. In this extreme case, the velocity potential ϕ is initially zero but rises to substantial magnitude in a very short time τ. During this process the first term in eqn (178), $\partial\phi/\partial t$, is so large as to dominate over all the others except $\rho^{-1}p_e$ itself. This gives the value

$$p_e = -\rho\partial\phi/\partial t \tag{180}$$

for the *transient pressure* appearing during the start up process. This result, clearly, is in full accord with eqn (173) for the potential of the flow generated at the end of that process.

Between these two extremes, the extended Bernoulli equation (eqn (178)) for unsteady irrotational flow describes a distribution of excess pressure in two parts: the *dynamic pressure* (a constant $-\frac{1}{2}\rho q^2$) given by eqn (179); and the *transient pressure* ($-\rho\partial\phi/\partial t$) given by eqn (180). This beautifully simple result, linearly combining the expressions for pressure in the two extreme cases, is of the greatest value in all irrotational-flow calculations.

To conclude this discussion we observe that the occurrence, which at first sight appears to be inconvenient, of an unknown function of the time on the right-hand side of the extended Bernoulli equation (eqn (178)) represents no uncertainty of any real importance since a uniform excess pressure $\rho f(t)$ present throughout the fluid is of such limited dynamical significance. Furthermore, the $f(t)$ term in eqn (178) can always be eliminated by taking

$$\phi - \rho \int_0^t f(t)\mathrm{d}t \qquad (181)$$

as a redefined form of the velocity potential; note that it necessarily possesses the characteristic property (in eqn (172)) of a velocity potential since, evidently, the gradient of expression (181) is equal to the velocity field **u** provided only that grad $\phi = \mathbf{u}$. In short, it is because the definition of the velocity potential leaves it arbitrary to within the addition of any function of time that this arbitrariness must be represented on the right-hand side of eqn (178).

6.2 Uniqueness of irrotational flow in simply connected regions

For the study of irrotational flows (that is, flows with curl $\mathbf{u} = 0$) the identification of a velocity potential ϕ satisfying grad $\phi = \mathbf{u}$ may be advantageous for many reasons: it allows the pressure field to be calculated from eqn (178) above; it simplifies the computation of the velocity field itself, as will later become clear; and its existence guarantees the uniqueness of that velocity field (given the values of its normal component at the boundaries of fluid), as we demonstrate in this section. First of all, however, we show that, if the region occupied by the irrotational flow of a fluid is of a certain type, called simply connected, then such a velocity potential necessarily exists.

The idea of the proof is to use Stokes's theorem (eqn (138)) to establish that for irrotational velocity fields **u** the line integral

$$\phi = \int_{\mathbf{x}_0}^{\mathbf{x}} \mathbf{u} \cdot \mathrm{d}\mathbf{x}, \qquad (182)$$

taken along a curved path within the fluid from a fixed point $\mathbf{x}_0$ to the general point $\mathbf{x} = (x, y, z)$, assumes an absolutely identical value whatever shape of path be chosen (Fig. 35). In fact, given any two such paths from $\mathbf{x}_0$ to $\mathbf{x}$, the difference between the value of the integral in eqn (182) taken along the two paths must be zero if the circulation (expression (136)) around the closed curve C (consisting of one of the paths followed by the other path taken backwards) is zero. This, by eqn (138), is necessarily the case for an irrotational motion (with zero vorticity $\boldsymbol{\omega}$) with one proviso: that *for any closed curve C within the fluid region there exists a surface S within the region which spans the closed curve C.* This is the property described by saying that the region is simply connected.

Fig. 35. Two alternative paths from $\mathbf{x}_0$ to $\mathbf{x}$ for the line integral of eqn (182) defining the velocity potential.

With the above proviso, then, the function ϕ defined as the line integral in eqn (182) is independent of path, and the difference between its values at the points $\mathbf{x} + d\mathbf{x}$ and $\mathbf{x}$ can be written as

$$d\phi = \left(\int_{\mathbf{x}_0}^{\mathbf{x}+d\mathbf{x}} - \int_{\mathbf{x}_0}^{\mathbf{x}} \right) \mathbf{u} \cdot d\mathbf{x} = \int_{\mathbf{x}}^{\mathbf{x}+d\mathbf{x}} \mathbf{u} \cdot d\mathbf{x} = \mathbf{u} \cdot d\mathbf{x}. \qquad (183)$$

We conclude that

$$\mathbf{u} = \operatorname{grad} \phi, \qquad (184)$$

since in fact, by eqn (183),

$$d\phi = u\,dx + v\,dy + w\,dz, \quad \text{giving } u = \frac{\partial \phi}{\partial x}, \quad v = \frac{\partial \phi}{\partial y}, w = \frac{\partial \phi}{\partial x}. \qquad (185)$$

This completes the proof that, for irrotational flow in a simply connected region, a velocity potential satisfying eqn (184) necessarily exists.

The condition for a region to be simply connected, which is printed above in italics, is a topological condition satisfied by fluid-filled regions in most, but not all, engineering applications. When, for example, a wing shape is being tested in a wind-tunnel (Fig. 36), the air around the wing model fills a simply connected region in case (a) with the wing attached to just one side of the tunnel, but not in case (b) with the wing attached to both sides of the tunnel. This is because in case (b) a closed curve such as C which embraces the wing cannot be spanned by any surface S lying within the region occupied by air; in fact, such a surface S must pass through the solid wing itself. By contrast, in case (a), a similar closed curve C can be spanned by a surface S passing around the wing-tip as shown, within the air. Later, in Section 6.4, we introduce the term *doubly connected* to describe a certain category of regions, including that of Fig. 36 (b), which are not simply connected; and we investigate their properties. In the meantime we concentrate, in the present section, on making further deductions from the existence of a velocity potential for irrotational flow in simply connected regions.

First, we demonstrate the uniqueness of irrotational flow of an incompressible fluid in a simply connected region of fluid L whose solid boundary is at rest. The boundary condition (eqn (73)) appropriate to the Euler model in

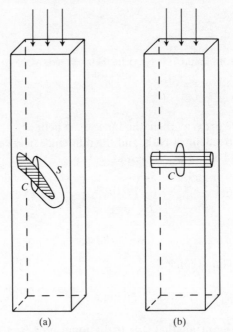

(a)   (b)

Fig. 36. Flows around different wing models in a wind tunnel are used to illustrate (a) a simply connected flow region and (b) a doubly connected flow region. A closed curve C passing around the model can be spanned by a surface S lying entirely within the flow in case (a) but not in case (b).

this case requires the vanishing of the normal component of fluid velocity ($\mathbf{u} \cdot \mathbf{n}$ = 0) on the boundary ∂L, from which it follows that the normal component of the vector $\phi\mathbf{u}$ (the velocity vector, *multiplied by the velocity potential*) is also zero:

$$(\phi\mathbf{u}) \cdot \mathbf{n} = 0 \text{ on } \partial L. \tag{186}$$

The Divergence Theorem (eqn (48)) applied to the vector $\phi\mathbf{u}$ therefore gives

$$\int_L \text{div}(\phi\mathbf{u}) \, dV = 0. \tag{187}$$

However, the integrand of eqn (187) can be expanded as

$$\text{div}(\phi\mathbf{u}) = \phi \, \text{div}\,\mathbf{u} + (\text{grad}\,\phi) \cdot \mathbf{u} = \mathbf{u} \cdot \mathbf{u}, \tag{188}$$

where the equation of continuity for an incompressible fluid,

$$\text{div}\,\mathbf{u} = 0, \tag{189}$$

has been used alongside eqn (184) for grad ϕ. Finally, then, eqn (187) becomes

$$\int_L (u^2 + v^2 + w^2) \, dV = 0, \tag{190}$$

which is possible for a continuous velocity field $\mathbf{u} = (u, v, w)$ only if the velocity is zero throughout L.

It is, of course, obvious that a motionless state of the fluid is irrotational; the result here proved, however, is that there is no *other* irrotational motion possible when the boundaries are instantaneously at rest.[†] This has a number of extremely important consequences, as follows.

We saw in Section 6.1 that flow started from rest by the impulsive motion of a solid boundary is initially irrotational. We can now show, however (provided the region of fluid is simply connected), that this initial irrotational motion *is the unique irrotational flow* compatible with the instantaneous motion of the solid boundary. For suppose that there were *two* irrotational velocity fields each satisfying the boundary condition (eqn (74)), stating that the normal component of fluid velocity $\mathbf{u} \cdot \mathbf{n}$ is equal to the normal component $\mathbf{u}_s \cdot \mathbf{n}$ of the instantaneous velocity of the solid boundary. In that case the *difference* of those two velocity fields would be an irrotational velocity field $\mathbf{u}$ with $\mathbf{u} \cdot \mathbf{n} = 0$ on the boundary, and we have just demonstrated that such a velocity field is everywhere zero. In short, the two velocity fields are necessarily identical.

For calculating an irrotational flow, given the normal component $\mathbf{u}_s \cdot \mathbf{n}$ of the instantaneous velocity of the solid boundary, it is of course of real practical use to know that any solution found must be the unique solution. In addition,

---

[†] We shall find this to be the key result that is true in simply connected, but false in doubly connected, regions (Section 6.4).

we have seen (Sections 5.4 and 5.5) how the unique solution with the normal components $\mathbf{u} \cdot \mathbf{n}$ and $\mathbf{u}_s \cdot \mathbf{n}$ of fluid velocity and boundary velocity matched together must in general have the tangential components of those velocities mismatched. This mismatch generates a vortex sheet at the surface, which grows into a boundary layer which (Section 5.6) may or may not separate so that free vorticity appears within the fluid, subsequently being convected and diffused.

From one standpoint, the fact that irrotational flow in simply connected regions is exclusively dependent on the *instantaneous* velocity of the boundary may be characterized by the statement that irrotational motion is entirely without memory. In other words, all of the 'memory' in a fluid flow lies in its vorticity; which, once generated, is subject as just mentioned to convection and diffusion.

One extreme case of this principle is evident when a solid boundary whose motion has generated a fluid flow is suddenly brought to rest. *If the fluid motion remained irrotational, the flow would instantaneously cease* since when the boundary is at rest the only irrotational flow is a state of zero motion (eqn (190)). Conversely, if any flow of fluid does continue, it must be a *rotational* flow associated with such vorticity as has been generated; and, in fact, associated in a rather immediate way which will be discussed in Section 6.3.

It may hardly seem necessary, perhaps, to emphasize that all the results given in this section apply to finite (that is, bounded) lumps L of fluid—those finite lumps, in fact, to which alone the Divergence Theorem (eqn (48)), used to derive eqn (187), is applicable. Bodies of fluid, after all, are necessarily finite! Nevertheless, it may often be helpful (as in Chapter 8) to model an extremely large lump of fluid by means of a theoretically *infinite* volume of fluid which possesses only an internal boundary, and where the fluid velocity tends to zero as the distance from that boundary increases without limit. On this subject, we at present merely mention one important feature of Chapter 8, namely that the conclusions on uniqueness there derived for such unbounded models are found to be exactly the same as those derived above for finite lumps L of fluid.

6.3 Related results for flows of given vorticity

From the uniqueness property for flows of zero vorticity (that is, irrotational flows) which was established in Section 6.2 there follows immediately a related uniqueness property for flows of *given* vorticity. Consider, indeed, two velocity fields $\mathbf{u}_1$ and $\mathbf{u}_2$ which, within a simply connected region L of fluid, have exactly the same distribution of vorticity $\boldsymbol{\omega}$:

$$\operatorname{curl} \mathbf{u}_1 = \operatorname{curl} \mathbf{u}_2 = \boldsymbol{\omega} \text{ in L.} \tag{191}$$

Then, for a given motion $\mathbf{u}_s$ of the solid boundary of L, eqn (74) represents the boundary condition satisfied by both flows

$$(\mathbf{u}_1 - \mathbf{u}_s) \cdot \mathbf{n} = 0 \text{ and } (\mathbf{u}_2 - \mathbf{u}_s) \cdot \mathbf{n} = 0 \text{ on } \partial \text{L.} \tag{192}$$

Equations (191) and (192) imply that the *difference* $\mathbf{u} = \mathbf{u}_1 - \mathbf{u}_2$ of those two velocity fields satisfies the very simple conditions

$$\text{curl } \mathbf{u} = 0 \text{ in L}, \mathbf{u} \cdot \mathbf{n} = 0 \text{ on } \partial L; \tag{193}$$

and these conditions, of course, were shown earlier, in eqns (186) to (190), to imply that the velocity field $\mathbf{u}$ is zero throughout L. In short, two such velocity fields $\mathbf{u}_1$ and $\mathbf{u}_2$, having the same vorticity field (eqn (191)) and satisfying the same boundary condition (eqn (192)), must actually be identical.

Thus there is just one velocity field $\mathbf{u}_1$ compatible with a given vorticity $\boldsymbol{\omega}$ and a given motion of the solid boundary. We next compare $\mathbf{u}_1$ to another velocity field $\mathbf{u}_0$ with the same distribution of vorticity

$$\text{curl } \mathbf{u}_0 = \text{curl } \mathbf{u}_1 = \boldsymbol{\omega} \tag{194}$$

but satisfying the boundary condition (eqn (73)) appropriate to a boundary instantaneously *at rest*

$$\mathbf{u}_0 \cdot \mathbf{n} = 0. \tag{195}$$

This time we cannot conclude that the difference $\mathbf{u}_1 - \mathbf{u}_0$ is zero. However, eqn (194) shows that $\mathbf{u}_1 - \mathbf{u}_0$ is an irrotational velocity field, while eqn (195) shows it to satisfy the boundary condition appropriate to motion of the boundary with velocity $\mathbf{u}_s$:

$$(\mathbf{u}_1 - \mathbf{u}_0) \cdot \mathbf{n} = \mathbf{u}_s \cdot \mathbf{n}. \tag{196}$$

We know, however (Section 6.2), that if L is simply connected there is one unique irrotational velocity field $\mathbf{u}_1 - \mathbf{u}_0$ satisfying this boundary condition, and that this is the gradient, grad ϕ, of a potential; thus

$$\mathbf{u}_1 = \text{grad } \phi + \mathbf{u}_0. \tag{197}$$

Equation (197) is a very useful representation of a flow with given vorticity which is compatible with a given motion of the boundary as being a linear combination of two velocity fields: (i) the irrotational flow grad ϕ compatible with the same boundary motion; and (ii) the velocity field $\mathbf{u}_0$ associated with the same vorticity distribution when the boundary is at rest. (Note that the velocity field (ii) represents the motion that would immediately be generated from the rotational velocity field $\mathbf{u}_1$ by bringing the boundary instantaneously to rest.)

Even though the representation in eqn (197) is a *linear* combination of the velocity fields (i) and (ii), while kinetic energy depends *quadratically* on velocity, we can nevertheless show that the kinetic energy of the combined velocity field $\mathbf{u}_1$ is the sum of the kinetic energies of the two fields (i) and (ii) separately. In fact, that kinetic energy is

$$\tfrac{1}{2}\rho \int_L (\mathbf{u}_1 \cdot \mathbf{u}_1) \, dV = \tfrac{1}{2}\rho \int_L (\text{grad } \phi + \mathbf{u}_0) \cdot (\text{grad } \phi + \mathbf{u}_0) \, dV$$

$$= \tfrac{1}{2}\rho \int_L (\text{grad } \phi)^2 \, dV + \rho \int_L (\mathbf{u}_0 \cdot \text{grad } \phi) \, dV + \tfrac{1}{2}\rho \int_L (\mathbf{u}_0 \cdot \mathbf{u}_0) \, dV, \tag{198}$$

where the first and last terms on the right-hand side represent the kinetic energies of the flow fields (i) and (ii) respectively. The middle term, however, is necessarily zero, as becomes evident when we write it as

$$\rho \int_{L} \text{div} (\phi \mathbf{u}_0) \, dV = \rho \int_{\partial L} \phi (\mathbf{u}_0 \cdot \mathbf{n}) \, dS = 0. \tag{199}$$

In this case, we have made use of the equation of continuity div $\mathbf{u}_0 = 0$ to write

$$\text{div} (\phi \mathbf{u}_0) = \phi \, \text{div} \, \mathbf{u}_0 + \mathbf{u}_0 \cdot \text{grad} \, \phi = \mathbf{u}_0 \cdot \text{grad} \, \phi, \tag{200}$$

and have then applied the boundary condition (eqn (195)).

This conclusion has a twofold interest. It identifies the kinetic energy of any rotational flow as being equal to the kinetic energy of an irrotational flow satisfying the same boundary conditions plus the kinetic energy of the flow due to the same vorticity distribution with the boundary at rest. Simultaneously, it demonstrates the Minimum Energy Theorem (due to Kelvin) which states that, of all flows satisfying given boundary conditions, the irrotational flow has least kinetic energy. This theorem gives us some help in understanding the special character of irrotational flows.

We conclude this section with just one illustration of the usefulness of knowing that there is a unique flow with given vorticity which satisfies given boundary conditions. Consider the effect of a plane solid boundary, at rest, on the fluid flow due to a nearby line vortex in the shape of a straight line parallel to the plane. Such a line vortex consists (Section 5.4) of a concentration of vorticity within only a very small distance of that line, and the circulation around any curve embracing the line once in a particular sense is the strength K of the line vortex. This definition leads to the characteristic velocity field illustrated in Fig. 30, representing a flow that is irrotational outside the line vortex itself. This, indeed, is one particular flow compatible with the above concentration of vorticity near the straight line; in addition, as explained in eqn (155), the tangential velocity on each circle in Fig. 30 represents for absolutely any flow compatible with that concentration of vorticity the *average* tangential velocity on such a circle within the fluid.

In the presence of a stationary plane boundary parallel to such a straight line vortex, it is evidently impossible for the velocity field to take the simple form shown in Fig. 10, a form which fails to satisfy the boundary condition $\mathbf{u} \cdot \mathbf{n} = 0$ on the nearby plane. However, the difference between the true velocity field and the velocity field of Fig. 30 (two fields with identical vorticity distributions) must be an irrotational flow,[†] and must furthermore be the unique irrotational flow whose distribution of normal velocity $\mathbf{u} \cdot \mathbf{n}$ will exactly cancel that associated with the velocity field of Fig. 30.

In order to identify the irrotational flow in question, we can use the idea

---

[†] As remarked in the footnote on p. 68.

(fruitful for dealing with plane boundary problems in many branches of physics) of the *mirror image*. Figure 37 shows: (a) that part of the velocity field of Fig. 30 which lies in the region of fluid limited by the plane boundary (dotted

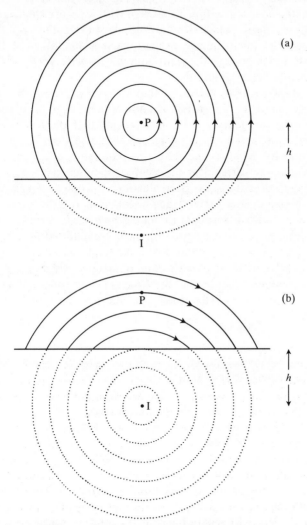

Fig. 37. Illustrating the method of images in the case of a straight line vortex P near a plane boundary in which the mirror image of the vortex is I.

(a) Motion around the vortex in the absence of the boundary with, however, dotted lines used for streamlines where they pass inside the boundary.

(b) The image motion, consisting of a line vortex placed at I with, again, dotted lines used for parts of streamlines inside the solid boundary.

The plain-line flows in (a) and (b) may be added to give the flow due to the straight line vortex in the presence of the plane boundary.

lines are used outside the region occupied by fluid); and (b) the mirror image of the same velocity field in the plane boundary. It is important to note that the flow field (b) is irrotational in the *whole* region occupied by fluid (that is, the region where full lines are used to designate its streamlines), as is evident from the fact that the flow field of Fig. 30 has zero vorticity everywhere except within the line vortex itself (which for flow (b) is outside the region of fluid). Furthermore, the mirror-image property makes the value of the normal velocity $\mathbf{u} \cdot \mathbf{n}$ at each point of the plane boundary *exactly equal and opposite* for the two velocity fields (a) and (b).

These last two statements mean that the velocity field (b) exhibits the properties required: it is an irrotational flow which, added to the velocity field (a), satisfies the boundary condition. Knowing that there can be only one unique flow with these properties, we can confidently use the linear combination of velocity fields (a) and (b) as the required flow associated with the presence of the line vortex near the solid plane boundary.

Incidentally, the analysis just given illustrates the power of a general method that is much used in later parts of the present book. It is the method by which *pieces* (that is, limited portions) of simple flow fields are used and, often, linearly combined in order to build up more complicated flow fields.

Line vortices do, of course, move with the fluid. Figure 37, illustrating the two velocity fields which together make up the flow field due to a line vortex near a plane boundary, shows that the velocity field (a) does not move the line vortex at all. However, the velocity field (b) does move it, parallel to the plane, at a speed

$$K/(4\pi h) \tag{201}$$

where h is the distance of the vortex from the plane. The combined flow field therefore moves the vortex at this speed. Note, too, that the *direction* of movement is precisely opposite to that which might have been expected by analogy with a rigid cylinder rolling along a plane!

The above model with extremely simple geometry leads to the interesting result that a line vortex near a solid surface tends to move parallel to that surface in a direction opposite to that of a rigid roller rotating in the same sense. Experiments show this to be a property of line vortices both when the geometry is as simple as in Fig. 37 and in much more general cases. The reader may test this by briefly moving the tip of a spoon through a cup of tea or coffee and then withdrawing it. Vorticity separating from the solid surface at the two edges of the spoon forms two line vortices which are clearly visible. After the spoon has been withdrawn, these vortices continue to move with the fluid, each vortex, in fact, being convected forwards by the velocity field of the other. However, when each vortex reaches the side of the cup, the effect analysed above becomes important; indeed, each vortex then acquires an additional movement parallel to the side of the cup which we observe to be such that the distance between them increases (Fig. 38).

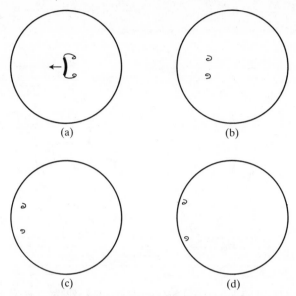

Fig. 38.   Vortices in a teacup, made visible by those dimples in the free surface that are generated by negative excess pressure at the centre of the vortex.
(a)   On being moved to the left, a teaspoon (represented by its intersection with the free surface) generates the vortices shown.
(b)   These persist after the teaspoon is withdrawn.
(c)   and (d). They move apart on reaching the teacup boundary.

6.4 Supplementary condition for uniqueness in doubly connected regions

All the results in Sections 6.2 and 6.3, whether on the existence of a velocity potential or on the uniqueness of irrotational flows or of flows with given vorticity, are critically dependent on the assumption that the region of fluid is simply connected. Any departure from that assumption, such as must occur when the region is traversed by a solid barrier as in figure 36(b), destroys the truth of all such results.

We can demonstrate this quickly by considering a cylindrical region of fluid penetrated by a central solid cylinder (Fig. 39). (The region is evidently not simply connected since the closed curve C_1 is not spanned by any surface S lying entirely within the region.) Within this region, uniqueness of irrotational flows can be disproved very readily by producing a wide range of different irrotational flows, *all* with the boundary at rest. We achieve this by using a method just referred to in rather general terms: the method depending on the use of just a *piece* of a simple flow field.

Once again, in fact, we use the flow field of Fig. 30 which we know to be

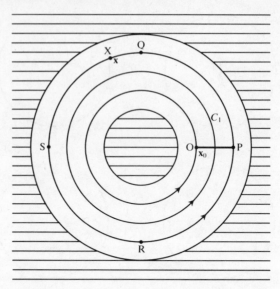

Fig. 39. Flow in a doubly connected region, constituting a counter-example not only to the uniqueness of irrotational flows but also to the existence of a velocity potential (eqn (182)) in such flow regions. The integral (eqn (182)) assumes different values when taken from x_0 to x along different paths OPQX and OPRSX.

irrotational everywhere except within the line vortex itself, so that it must be irrotational everywhere within the region of fluid illustrated in Fig. 39. This excludes the whole volume occupied by the central solid cylinder, *including its axis* (where the line vortex lies). At the boundaries, furthermore, the velocity field is directed tangentially, so that the boundary condition $u \cdot n = 0$ is satisfied.

Therefore, there is not just one unique irrotational flow which satisfies the boundary condition, i.e. that *stationary* state (with the fluid at rest) which is the only possibility in a simply connected region. Instead, there is a wide range of such flows, including the flow illustrated in Fig. 39 for *all* values of the strength K (among these, the case $K = 0$ describes the stationary state).

We show next how the irrotational flow in the doubly connected region with its boundary at rest illustrated in Fig. 39 is an example contrary not only to uniqueness but also to the existence of a velocity potential. If, in fact, we attempt to construct a velocity potential by evaluating the integral in eqn (182) along the two different paths from x_0 to x shown in Fig. 39, each starting along a radius through x_0 and then continuing tangentially to the circle through x, we obtain different values of the integral for each path. There is, of course, no contribution to the integral in eqn (182) from the radial portion OP of either path, where the radial displacement dx is perpendicular to the tangential velocity u. However, in the tangential portion PQX, the velocity u of magnitude

$K/(2\pi r)$ is everywhere parallel to a displacement which in total amounts to $r\theta$ (where θ is the angle between the radius through $\mathbf{x_0}$ and the radius through $\mathbf{x}$). This gives a contribution $K\theta/(2\pi)$ to the integral

$$\int_{OPQX} \mathbf{u} \cdot d\mathbf{x} = K\theta/(2\pi). \qquad (202)$$

On the alternative tangential part of the path PRSX the velocity $K/(2\pi r)$ is directed oppositely to a displacement which in total amounts to $r(2\pi - \theta)$, thus giving a contribution $K(\theta - 2\pi)/(2\pi)$ to the integral

$$\int_{OPRSX} \mathbf{u} \cdot d\mathbf{x} = -K + K\theta/(2\pi). \qquad (203)$$

Figure 39 represents a flow, then, where the integral in eqn (182) from $\mathbf{x_0}$ to $\mathbf{x}$ takes different values according to the path chosen. As might have been expected, the difference between the above values (eqns (202) and (203)) is the circulation

$$\int_{PQSRP} \mathbf{u} \cdot d\mathbf{x} = K \qquad (204)$$

around the central cylinder, a circulation which is not obliged to be zero since no surface lying within the region of irrotational flow spans the closed curve in question.[†]

There are, in fact, other paths from $\mathbf{x_0}$ to $\mathbf{x}$ which yield different values for the integral in eqn (182) from *either* eqn (202) *or* eqn (203). For example, a path from O to P, followed by n complete anticlockwise circuits PQSRP, followed by the direct path PQX would give a value $nK + K\theta/(2\pi)$ to the integral, and negative values of n here could represent the possibility of the said complete circuits being clockwise.

We are ready now to explain the description *doubly connected* which is applied to regions topologically equivalent to that of Fig. 39. From the standpoint of irrotational-flow theory they are regions where the circulation around *any* given closed curve C has a value numerically related to the circulation K around one particular curve C_1. For example, when the given closed curve C is of a shape capable of being stretched or otherwise deformed entirely within the region until it coincides with C_1, then the ensemble of successive positions of the curve during deformation make up a surface spanning C and C_1 (and still in the region) to which Stokes's theorem can be applied to show that the circulation around C coincides with the circulation K around C_1. More generally, however, a closed curve such as C may embrace a

[†] It should also be noted that, besides giving different results, *neither* eqn (202) *nor* eqn (203) specifies a single continuous function ϕ whose gradient could be equated to the velocity field, essentially, because of the well known difficulty that no single-valued definition of the angle θ can avoid a discontinuity (e.g. between the values $\theta = 0$ and $\theta = 2\pi$).

central body n times, in which case it may be capable of being deformed into C_1 *not* just as described once but as described n times, so that the circulation around C is equal to nK.

A region is said to be doubly connected if it contains one specially identified closed curve C_1 with the property that every closed curve C in the region can be deformed, entirely within the region, into the closed curve C_1 taken n times, where n is some integer (positive, negative or zero). For irrotational flow within the region, this means that the circulation around C takes the value nK, where K is the circulation around C_1. For a doubly connected region, in short, the circulation K around the specially identified closed curve C_1 determines the circulation around each and every other closed curve C.

Actually, the value of this parameter K (the circulation around the specially identified closed curve C_1) does need to be prescribed as a *supplementary condition* if the uniqueness of irrotational flows for given normal velocity $\mathbf{u}_s \cdot \mathbf{n}$ of the solid boundary (demonstrated in Section 6.2 for simply connected regions) is to be extended to doubly connected regions. Suppose, in fact, that there were two irrotational velocity fields $\mathbf{u}_1$ and $\mathbf{u}_2$ with the same value of K. Then the difference velocity field $\mathbf{u} = \mathbf{u}_1 - \mathbf{u}_2$ would not only have $K = 0$ but would also have zero circulation (equal to nK on the above analysis) around *any closed curve* C in the region. Therefore the integral in eqn (182) is independent of path and it accordingly satisfies eqn (184). Finally, eqns (186) to (190) lead to the previous conclusion that the velocity field $\mathbf{u}$ is everywhere zero. This demonstrates the uniqueness property: two velocity fields $\mathbf{u}_1$ and $\mathbf{u}_2$ satisfying the same boundary conditions (on the normal component of velocity at the boundary) *and the same supplementary condition* on the value of the circulation K are necessarily identical.

This variant on the conditions for uniqueness of irrotational flows is by no means of purely theoretical interest. Its great practical significance becomes clear later, in Chapter 10 on flows with circulation in which, furthermore, detailed consideration is given to those conditions which may appropriately be used to fix the (otherwise arbitrary) value of K in real flows around shapes of, for example, 'aerofoil' type.

7
Three-dimensional examples of irrotational flows

It is above all the fact that irrotational flows are governed by a linear equation (eqn (151)), rather than by the general nonlinear momentum equation (eqn (68)) of the Euler model, which (relatively speaking) makes them so much easier to calculate. Different irrotational velocity fields, because they all satisfy the same linear equation, may be linearly combined (in various proportions) so as to generate the unique irrotational flow satisfying particular conditions. Figure 37 illustrates the effectiveness of this approach. It exemplifies, too, the value of using *pieces* (that is, limited portions) of simple flow fields and linearly combining them to build up more complicated flow fields. It is noteworthy that, even if a particular flow field like that of Fig. 37 (b) possesses a 'singularity' (the line vortex, where the smooth irrotational character of the flow breaks down), nevertheless a piece of the flow excluding that singularity may be employed as the added irrotational flow which needs to be incorporated if the boundary condition is to be satisfied.

The present chapter includes several further illustrations of all of the above ideas. It also applies the criteria for boundary layer separation (Section 5.6) to each calculated irrotational flow around a solid body in order to estimate whether it can in practice be used to represent a real flow.

Only elementary mathematical analysis (combining at most three simple flow fields) is used in the present chapter. There follows, in Chapter 8, a more comprehensive analysis of a wide range of cases. Understanding of the analytical material in these chapters is an important background to appreciating modern methods for computing irrotational flows, in order both to understand the types of numerical analysis used and to be able to judge whether the flow calculated is capable of remaining irrotational. Indeed, while the determination of three-dimensional irrotational flows has gradually become more straightforward owing to the widespread availability of the necessary computer power, relatively subtle skills continue to be needed for the effective utilization of such calculations.

7.1 Laplace's equation in three dimensions

In the simply connected regions of fluid with which Chapters 7 and 8 are concerned, the velocity field can necessarily be written (Section 6.2) as the gradient

$$\mathbf{u} = \text{grad}\,\phi = \nabla\phi \tag{205}$$

of a velocity potential ϕ. By the definition (expression (28)) of the gradient operator, grad ϕ is everywhere directed along the normals **n** to the surfaces ϕ = constant (often called *equipotential* surfaces) in the sense ϕ increasing, its magnitude being $\partial\phi/\partial n$. Thus, the streamlines of the velocity field **u** must everywhere cross the equipotential surfaces at right angles. Furthermore, the fluid speed $\partial\phi/\partial n$ must be greater wherever two neighbouring equipotential surfaces come closer to each other.

The last result is reminiscent of the solenoidal property (eqn (41)) of the Euler model, that along a streamtube the fluid speed q becomes greater where the streamtube area S is reduced. Irrotational flows possess both properties, being characterized by a set of equipotential surfaces and by a set of streamlines crossing them at right angles (often called their *orthogonal trajectories*), and with fluid speed increasing both wherever neighbouring equipotential surfaces come closer together and wherever the streamtube area is constricted (Fig. 40).

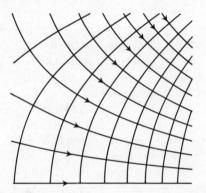

Fig. 40. Streamlines (marked with arrows) and equipotential surfaces (crossing them at right angles) within a certain region of a particular irrotational flow. Where the streamlines come closer together, indicating greater fluid speed, the equipotential surfaces (plotted for a set of equally spaced constant values of ϕ) are also closer together.

Analytically, the solenoidal property is characterized by the equation of continuity

$$\text{div}\,\mathbf{u} = \nabla\cdot\mathbf{u} = 0 \tag{206}$$

for the velocity field **u** of an incompressible fluid. Applied to the irrotational velocity field (eqn (205)), eqn (206) becomes the famous Laplace's equation, which necessarily is satisfied by the velocity potential on the Euler model:

$$\nabla^2\phi = 0, \tag{207}$$

where div grad ϕ or $\nabla\cdot\nabla\phi$ has been shortened to $\nabla^2\phi$. Written out in full,

Laplace's equation (eqn (207)) becomes

$$\partial^2\phi/\partial x^2 + \partial^2\phi/\partial y^2 + \partial^2\phi/\partial z^2 = 0. \tag{208}$$

The problem of calculating an irrotational flow is a problem of finding that solution of Laplace's equation (eqn (207)) which satisfies the appropriate boundary condition. At a solid surface, for example, the boundary condition (eqn (74)) becomes a condition

$$\partial\phi/\partial n = \mathbf{u}_s \cdot \mathbf{n} \tag{209}$$

on the component $\partial\phi/\partial n$ of grad ϕ normal to the solid surface.[†]

Laplace's equation arises similarly in other branches of physics and, in particular, is often satisfied by the potentials of conservative fields of force (for example by the electrostatic potential in a region without electric charge). Some methods of obtaining solutions devised for other applications may evidently be used to determine irrotational flows, although we often find that the types of solution in which we are interested are rather different in theoretical fluid dynamics.

Chapters 7 and 8 are devoted to irrotational flows in which all three terms in eqn (208) are of comparable importance. Thus, we postpone until Chapter 9 any consideration of the special methods of treatment that are available in cases in which, with appropriate choice of the coordinate axes, the last term in eqn (208) is of negligible importance compared with the other two. Then, to good approximation, ϕ satisfies the two-dimensional Laplace equation $\partial^2\phi/\partial x^2 + \partial^2\phi/\partial y^2 = 0$ for which a particularly convenient special theory exists. It is, however, convenient to postpone the use of that special theory to study the characteristic properties of 'two-dimensional' irrotational flows until they can be investigated against the general background of knowledge of fully three-dimensional irrotational flows which will be obtained from Chapters 7 and 8.

7.2 Spherically symmetrical motions

Often in a very large region of fluid a process is occurring whereby fluid is continuously being abstracted into a small opening and thence removed from the main body of fluid. Figure 41(a) illustrates such a case where the size of the region of fluid as a whole is enormously greater than the size of the small orifice through which fluid is removed. For comparison, Fig. 41(b) shows a convenient and widely used mathematical model of such a process: the 'point sink', where fluid is deemed to be removed at a certain rate J (in units of fluid volume per unit time) while the dimensions of this location are taken as being reduced

---

[†] This use of $\partial\phi/\partial n$ as a derivative normal to the solid surface must, of course, not be confused with its use earlier in this section as a derivative normal to the equipotential surface.

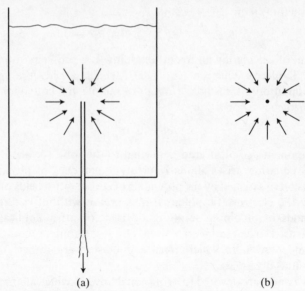

Fig. 41. This illustrates the 'point sink' by means of: (a) the continuous removal of fluid from a large tank through a narrow pipe; and (b) the standard idealized mathematical model.

to zero. The mathematical model then exhibits a 'singularity' (the point at its centre where fluid is considered to be *disappearing* at a rate J).

The flow into a point sink is spherically symmetric, since removal of fluid at the point draws more fluid inwards equally from all directions. The same general tendency is found, furthermore, to characterize a flow into a small orifice as in Fig. 41(a). Except very near the orifice itself, the abstraction of fluid there tends to induce inflow towards the orifice from all directions equally.

The spherically symmetric flow into a point sink is evidently without rotation; it is an irrotational flow with the streamlines directed radially inwards towards the point. The equipotential surfaces (which the streamlines must cross at right angles) are therefore spherical surfaces. If

$$r = (x^2 + y^2 + z^2)^{\frac{1}{2}} \tag{210}$$

represents distance from the point sink (taken as the origin of the coordinates), the velocity of inflow across the spherical surface at distance r, with area $4\pi r^2$, is necessarily

$$J/(4\pi r^2) \tag{211}$$

if the volume flow across the surface is to balance the rate J of removal of fluid at the sink. The velocity potential ϕ must therefore take the value

$$\phi = J/(4\pi r) \tag{212}$$

in order that the gradient of ϕ should have the magnitude in expression (211) and be directed radially *inward* (along the direction ϕ increasing).

In the original illustrative example of Fig. 41(a), we may test now whether the irrotational model of the flow external to the pipe is one that can persist or whether it is necessarily modified as a result of boundary layer separation. We see, in fact, that the flow along the solid external surface of the extract pipe, with velocity given by expression (211) where r is distance from the orifice, is an *accelerating* flow. Therefore, there is no tendency to separation (Section 5.6), and the irrotational flow can persist.

Similarly, if we utilize just a *piece* (in fact, the upper half) of the irrotational flow field of Fig. 41(b) to represent outflow from a small opening in the flat bottom of a large vessel (Fig. 42), the inflow along the bottom towards that opening is again an accelerating flow, with velocity given by expression (211). Once more, then, the 'sink' flow is readily realizable, showing no tendency to boundary layer separation.

We may readily check that the most general spherically symmetrically solution of Laplace's equation (eqn (208)) can differ from eqn (212) only by an unimportant constant. Thus, if ϕ is a function $f(r)$, with r defined by eqn (210), then we calculate that $\partial \phi / \partial x = x r^{-1} f'(r)$ and thence that

$$\partial^2 \phi / \partial x^2 = x^2 r^{-2} f''(r) + (r^{-1} - x^2 r^{-3}) f'(r), \tag{213}$$

and similarly with the other terms in eqn (208) which therefore becomes

$$f''(r) + 2r^{-1} f'(r) = 0, \tag{214}$$

of which the general solution takes the form $f = A + Br^{-1}$ (with A and B constant). The added constant A in a potential ϕ makes no difference to its velocity field grad ϕ, so it can be ignored. The constant B, if positive, gives the point sink flow with $B = J/4\pi$, but we must briefly consider the alternative possibility, with B negative.

Such brief consideration (precisely in accordance with principles already explained) leads to a conclusion which at first sight may completely astonish anyone familiar only with branches of physics where the governing equations are linear. Admittedly, we have emphasized that the equations governing irrotational flow are linear (and, in particular, Laplace's equation (eqn (207)) is linear) so that if we change the sign of the solution (eqn (212)) of Laplace's equation we obtain an irrotational flow: the so-called *point source*

$$\phi = -J/(4\pi r), \tag{215}$$

identical with that of Fig. 41(b) but with the direction of flow reversed on each streamline. We shall, indeed, make use of this irrotational flow field (eqn (215)), with a point singularity at which new fluid is appearing at the rate J units of volume per unit time (whence its description as a 'point source'); but we shall use it (for example in Section 7.3) as just a convenient solution of Laplace's equation. The question of whether the irrotational flow in question can persist as a real motion with solid boundaries depends on rules (Section 5.6) which are derived from the real, nonlinear equations of motion of a fluid. There is therefore no reason whatsoever why the reversal of flow direction in either Fig. 41(a) or Fig. 42 should produce an irrotational flow that is capable of persisting.

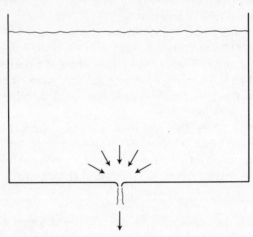

Fig. 42.   The upper half of the 'point sink' flow of Fig. 41(b) may be used to model outflow from a large tank through a small opening in its flat bottom.

Simple everyday experience tells us, in fact, that fluid issuing from a pipe into a fluid-filled region with the geometry of Fig. 41(a) or Fig. 42 shows no tendency whatsoever to follow the streamlines of those figures in reverse and spread out in all directions. Instead, the fluid issuing from the pipe forms a *jet*; that is, it broadly maintains its original direction with a gradually widening, eddying motion (Fig. 43).

In the meantime, the criteria of Section 5.6 already indicate unambiguously that the irrotational motions obtained by reversing the direction of flow on each streamline in Fig. 41(a) and 42 would undergo a massive boundary layer separation at the orifice itself. This is because, along the solid surface, each irrotational flow (eqn (215)) would involve a radially outward motion of speed $J/(4\pi r^2)$ which decreases extremely abruptly as r increases. The consequent boundary layer separation at the orifice generates the shedding of eddies that produces the jet-type flow of Fig. 43.

The methods we have described do not, of course, indicate the precise character of that jet-type flow. They show merely that, of two irrotational flows (eqns (212) and (215)) which differ only in the *sign* of the velocity

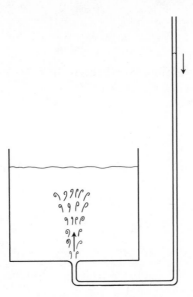

Fig. 43. The motion in Fig. 42 is *not* 'reversible'. In fact, flow into a large tank through a small opening in its flat bottom fails even approximately to follow in reverse the streamlines of Fig. 42; rather, it separates from the boundary and forms a turbulent jet.

potential, one (the point sink) can be easily realised in a flow with quite simple geometry, while the opposite flow (the point source) is not at all easily realised.

Before giving further illustrations of the methods for assessing the value and applicability of different irrotational flow fields in the presence of solid boundaries, we may illustrate the use of a *piece* of the spherically symmetric flow field (eqn (212)) to represent an interesting motion without solid boundaries. We give the classical analysis of the collapse of a spherical cavity in a liquid which is produced by a positive pressure P of the liquid in the general neighbourhood of the cavity.[†] The hydrostatic component in the pressure distribution, which leads to the buoyancy force causing the bubble to rise slowly, is however neglected on the ground that the speeds of collapse are found to be so much greater. It is, of course, this neglect which allows the mathematical model to be spherically symmetric.

If at time t the radius of the cavity is a decreasing function $a(t)$ then the instantaneous flow of liquid for $r \geqslant a$ may be modelled as a piece of the point

[†] We have explained in the footnote on p. 15 the tendency of liquids to 'cavitate' (that is, to form an assemblage of small bubbles) if the pressure actually reaches *negative* values. One of the reasons why engineering design usually seeks to avoid cavitation is associated with the abrupt collapse of such a bubble when it moves into a region of positive pressure P. The mathematical model of collapse given in the text neglects, however, the presence of other bubbles, and also neglects surface-tension effects together with the pressure of any vapour within the cavity.

sink flow of Fig. 41(b) with

$$J = -4\pi a^2 \dot{a} \tag{216}$$

as the rate of decrease of volume of the cavity (Fig. 44). By eqn (212), this makes the velocity potential

$$\phi = -a^2 \dot{a} r^{-1}, \tag{217}$$

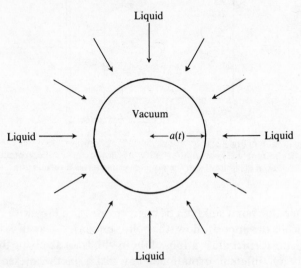

Fig. 44. Inward liquid motion outside a collapsing spherical cavity whose radius $a(t)$ is diminishing at an increasing rate.

so that the velocity of liquid radially inwards is

$$q = -\partial\phi/\partial r = -a^2 \dot{a} r^{-2}, \tag{218}$$

which satisfies the necessary boundary condition at $r = a$ where it evidently coincides with the velocity $(-\dot{a})$ of the cavity boundary.

The corresponding distribution of the pressure p is given by eqn (178) where, however, p_e is to be replaced by p because the hydrostatic pressure distribution is neglected. From eqns (217) and (218) we obtain

$$-(2a\dot{a}^2 + a^2\ddot{a})r^{-1} + \tfrac{1}{2}a^4\dot{a}^2 r^{-4} + \rho^{-1}p = \rho^{-1}P, \tag{219}$$

where the arbitrary function of time on the right-hand side of eqn (178) has been replaced by $\rho^{-1}P$ in order that the pressure p shall take the value P at a large distance from the cavity. Equation (219) specifies the pressure distribution as a sum of 'transient' and 'dynamic' pressures (represented by the first and second terms respectively).

At the cavity boundary $r = a$, however, the pressure p in the liquid must match the zero pressure within the cavity. By eqn (219), this gives

$$-\tfrac{3}{2}\dot{a}^2 - a\ddot{a} = \rho^{-1}P, \tag{220}$$

which incidentally means that the pressure distribution given by eqn (219) can conveniently be rewritten

$$p = P(1 - ar^{-1}) + \tfrac{1}{2}\rho\dot{a}^2(ar^{-1} - a^4r^{-4}). \tag{221}$$

Equation (220) is a differential equation for the cavity radius $a(t)$. It may be integrated once, after we have multiplied both sides by the integrating factor $(-4\pi\rho a^2\dot{a})$, to give

$$2\pi\rho a^3\dot{a}^2 = \tfrac{4}{3}\pi(a_0^3 - a^3)P, \tag{222}$$

where a_0 is the radius of the cavity when $\dot{a} = 0$ (that is, before collapse begins). Equation (222) has indeed a simple physical interpretation: it equates the total kinetic energy in the velocity field in eqn (218) for $r \geqslant a$, namely

$$\int_a^\infty (\tfrac{1}{2}\rho q^2)4\pi r^2 dr = 2\pi\rho\int_a^\infty a^4\dot{a}^2 r^{-2} dr = 2\pi\rho a^3\dot{a}^2, \tag{223}$$

to the work done by a pressure difference P as it reduces the cavity volume from $\tfrac{4}{3}\pi a_0^3$ to $\tfrac{4}{3}\pi a^3$.

The solution of eqn (222) is easily computed, and is plotted in Fig. 45. The abscissa is t/t_0, where $t_0 = a_0/q_0$ is the time taken to travel a distance a_0 (the initial radius of the cavity) at a speed q_0 defined[†] as $(P/\rho)^{\frac{1}{2}}$. The ordinates are a/a_0 (the cavity radius related to a_0) and $(-\dot{a}/q_0)$ (the speed of contraction related to q_0). Note that collapse is becoming complete ($a/a_0 \to 0$ and $-\dot{a}/q_0 \to \infty$) as t/t_0 approaches 0.915.

In the meantime, eqn (221) shows that the pressure p has nowhere become negative. On the contrary, the transient pressure effects during the later stages of collapse generate positive pressures in excess of P itself. The curve p_{max}/P in Fig. 45 demonstrates that cavity collapse under pressure P can temporarily produce much greater peak pressures p_{max} in the midst of the fluid. Here then is a first crude mathematical model which gives a preliminary understanding of why cavity formation and later abrupt collapse may be undesirable flow features, responsible for enhanced local loading such as can cause damage to solid surfaces.

[†] Since this speed q_0 is already $1\,\mathrm{m\,s^{-1}}$ at such a modest pressure P as $1000\,\mathrm{N\,m^{-2}}$, the processes described in Fig. 45 are fast processes (thus $t_0 = 0.01\mathrm{s}$ if $a_0 = 0.01\,\mathrm{m}$ and $q_0 = 1\,\mathrm{m\,s^{-1}}$).

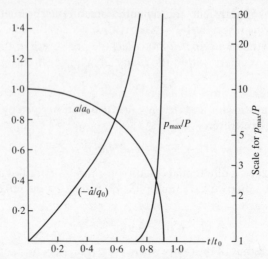

Fig. 45. Collapse of a cavity, initially at rest with radius a_0, in liquid at pressure P. The cavity radius a is plotted as a function of the elapsed time t divided by $t_0 = a_0/q_0$, where $q_0 = (P/\rho)^{\frac{1}{2}}$. The nondimensional rate of decrease of radius $(-\dot{a}/q_0)$ is also plotted. Finally, the maximum pressure p_{max} in the liquid, which remains equal to P for $t/t_0 < 0.73$, is shown to rise to very much greater values (see right-hand scale) as the instant of collapse is approached.

7.3 Axisymmetrical fairings

Returning next to steady flows, while however moving away from spherical symmetry, we now pursue further the theme of using just a piece of a flow. As before the piece used excludes that flow's singularity, but the theme is now extended to a case in which the precise piece to be used cannot be specified in advance.

The method, indeed, starts by considering in a purely theoretical way the particular irrotational flow obtained by the linear combination of two very simple flows: (i) the point source flow with velocity potential in eqn (215) and (ii) a completely uniform flow with constant velocity U parallel to the z axis, which evidently has velocity potential

$$\phi = Uz. \tag{224}$$

Although it is impossible for this combined flow to be realised in practice, we shall find that a particular piece of it really is useful and does satisfy the conditions needed to avoid boundary layer separation.

Taking the origin of coordinates to be at the point source itself as in Section 7.2, we can write the combined velocity potential as

$$\phi = Uz - J/(4\pi r), \tag{225}$$

with r defined as in eqn (210). This potential is symmetrical about the z axis and can be expressed entirely in terms of two *cylindrical coordinates*: z itself and the distance s from the z axis, where

$$s = (x^2 + y^2)^{\frac{1}{2}}, \quad r = (s^2 + z^2)^{\frac{1}{2}}. \tag{226}$$

At every point the flow velocity associated with eqn (225) is obtained by vector addition of a uniform velocity U in the z direction and a velocity of magnitude $J/(4\pi r^2)$ directed radially outward from the origin. For small r, the latter motion (due to the point source) is dominant; for large r, the flow becomes almost a uniform stream. At one point S (the stagnation point) the two velocities exactly cancel; this point is on the axis at a distance $r = (J/4\pi U)^{\frac{1}{2}}$ upstream of the point source (Fig. 46).

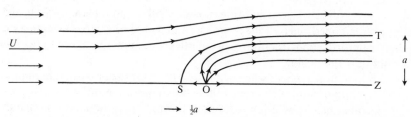

Fig. 46. The flow (eqn (225)) obtained by combining (i) a point source with (ii) a uniform flow. The whole motion is axisymmetric about the z axis SOZ. At the stagnation point S the two motions exactly cancel, and the flow divides to follow the dividing streamline ST; or, more correctly, a dividing streamtube, obtained by rotating ST about the axis of symmetry SOZ.

On the axis near the point S, the velocity is directed *towards* S both upstream (where the uniform flow velocity exceeds that due to the point source) and downstream (where the opposite is the case). Off the axis near S, the velocity is directed *away* from S, since the uniform flow cannot cancel the component of the point-source flow perpendicular to the z axis. Thus there is a *dividing streamline* phenomenon at S: flow on the axis towards S from both sides divides at S and moves away from the axis, initially at right angles. As the dividing streamline moves further from the axis, the z component of velocity due to the point source becomes weaker than the uniform flow velocity U, thus allowing the streamline to turn around towards the positive z direction. Far downstream, the uniform flow velocity is dominant and the dividing streamline has become parallel to the z axis.

Figure 46 shows this behaviour of the dividing streamline in a meridian section through the z axis. The whole axisymmetrical flow is obtained by rotating the figure about the z axis, so that the dividing streamline constitutes a *streamtube* representing the division between fluid arriving from far upstream and fluid emerging from the point source. Within this streamtube, J units of volume per unit time are generated at the source and, much later, become directed downstream at velocity U in a cylindrical region of radius a, within which the volume flow is $\pi a^2 \, U$. Therefore

$$J = \pi a^2 U, \quad a = (J/\pi U)^{\frac{1}{2}}. \tag{227}$$

Note that the distance of the stagnation point S upstream of the origin, previously determined as $(J/4\pi U)^{\frac{1}{2}}$, may now be written as $\frac{1}{2}a$ in terms of the downstream radius a of the streamtube.

For a class of theoretical irrotational flows including that of Fig. 46, the fruitful idea introduced by William Rankine (1820–72) was to utilize *only* the piece of flow outside the dividing streamtube; that is, the piece occupied by fluid arriving from far upstream. This piece of the flow of Fig. 46 can then be regarded as the steady irrotational flow around a solid body of revolution in the shape of the dividing streamtube. If that irrotational flow can persist, then the shape of solid surface in question is, as we shall see, suitable as a *fairing*, that is as a 'streamlined' surface shape which can be used to enclose machinery or other engineering constructions so as to offer minimum resistance to the flow of fluid past the fairing.

Extending the argument which led to eqn (227) for the downstream radius a of the dividing streamtube, we can obtain the shape of the entire streamtube in Fig. 46 (sometimes known as 'the Rankine fairing') from the condition that the volume flow through it is exactly J. Suppose, for example, that, instead of considering the volume flow through the downstream disc of radius a, we consider just an inner disc of radius s within it together with the surface of a cylinder of radius s extending from the disc to where it meets the dividing streamtube at a point with cylindrical coordinates s, z (Fig. 47). A total of J units of volume must cross this surface per unit time. The uniform stream generates a volume flow

$$\pi s^2 U \tag{228}$$

across the disc (which is taken too far downstream for the point source flow to be significant there). The point source generates a volume flow

$$\tfrac{1}{2}J(1 - zr^{-1}) \tag{229}$$

across the cylindrical surface. This may be obtained from the velocity potential of eqn (215) of the point source flow, with $r^2 = s^2 + z^2$, as

$$\int_z^\infty (\partial\phi/\partial s)2\pi s\,\mathrm{d}z = \tfrac{1}{2}J \int_z^\infty s^2 (s^2 + z^2)^{-\frac{3}{2}}\,\mathrm{d}z, \tag{230}$$

which takes the value in expression (229). It may, alternatively, be derived from the total volume flow J emerging from the source equally in all directions by considering the *solid angle* $2\pi(1 - zr^{-1})$ subtended at the source by the cylindrical surface in Fig. 47 as a proportion of the complete solid angle 4π.

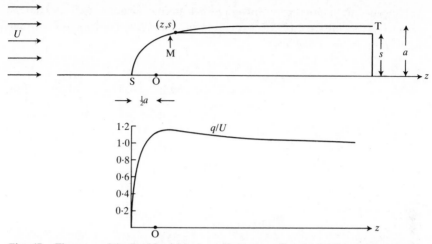

Fig. 47. Flow around the 'Rankine fairing', specified by regarding the dividing streamtube ST in Fig. 46 as a solid boundary. Its shape is derived from the condition that the volume flow from the source at O across a particular surface, consisting of a downstream disc of radius s extended upstream by a cylinder of radius s to where it meets ST at the point (z, s), must be exactly J. The lower diagram plots the fluid speed q on the surface (just outside the boundary layer) as a function of the coordinate z. It reaches its maximum value of 1.15 U at the point M.

Finally, the total volume flow J is necessarily equal to the sum of expressions (228) and (229):

$$J = \pi s^2 U + \tfrac{1}{2} J \left(1 - z r^{-1}\right). \tag{231}$$

With eqns (227) used to express J in terms of the downstream radius a of the dividing streamtube, eqn (231) may be rewritten

$$z (s^2 + z^2)^{-\frac{1}{2}} = 2 s^2 a^{-2} - 1, \tag{232}$$

which is readily solved for z to give the equation of 'the Rankine fairing' as

$$z = (s^2 - \tfrac{1}{2} a^2)(a^2 - s^2)^{-\frac{1}{2}}. \tag{233}$$

This is the equation which was used to plot its shape in Fig. 46.

It is similarly possible to obtain the equation for *any* axisymmetrical streamtube in the flow outside the Rankine fairing in Fig. 46 simply by using eqn (231) with the left-hand side (the constant volume flow within the dividing streamtube) replaced by any constant volume flow *greater* than J. In fact, values 1.5 J and 2J were used[†] to plot the external streamtubes shown (in meridian section) in Fig. 46.

[†] The technique outlined, whereby axisymmetric streamtubes are plotted by using the solenoidal property that the volume flow through any such streamtube takes a constant value, is often set out, however, using a different nomenclature (that of *Stokes's stream function*, defined as volume flow divided by 2π).

The fluid speed q in the flow around the Rankine fairing is derived from the velocity potential (eqn (225)), expressed in cylindrical coordinates through eqns (226), by the equation

$$q^2 = (\partial\phi/\partial z)^2 + (\partial\phi/\partial s)^2$$
$$= U^2 + UJz/(2\pi r^3) + J^2/(16\pi^2 r^4). \tag{234}$$

We are particularly interested in the distribution of fluid speed on the body surface itself where (i) the value of z/r is given by eqn (232), and (ii) by squaring eqn (232) and subtracting it from unity we obtain

$$r^{-2} = 4a^{-2}(1 - s^2 a^{-2}). \tag{235}$$

Using these results in eqn (234), with J replaced by $\pi a^2 U$ from eqns (227), we obtain

$$q^2 = U^2(4s^2 a^{-2} - 3s^4 a^{-4}). \tag{236}$$

Equation (236) is used in Fig. 47 to plot q/U along the surface of the body as a function of the axial coordinate z (which in turn is related to s by eqn (233), of course).

Figure 47 shows that the oncoming stream of velocity U is brought to rest at S by the presence of the solid body, and then accelerates until at M the speed q takes its maximum value $1.15U$. Thereafter, there is a very gradual retardation from this maximum value to the original velocity U of the undisturbed stream.

It might have been expected from the solenoidal property, of course, that the compression of streamtubes resulting from the insertion of a solid body in the stream would raise the fluid speed q to values greater than those in the oncoming flow. On the other hand, the importance of smoothly shaped fairings like that of Fig. 47 is that they keep the maximum velocity excess down to a modest percentage (here, 15 %) and allow the subsequent retardation to be only very gradual.

We have seen (Section 5.6) that separation of the boundary layer is avoided on surfaces where the flow outside the boundary layer is only weakly retarded as in this case. Therefore, we can expect (correctly) that the irrotational flow around the Rankine fairing illustrated in Figs. 46 and 47 should be realized to good approximation outside a thin boundary layer. Some important further consequences of this conclusion are given in the next section.

7.4 Drag of streamlined bodies

The general popular use of the adjective 'streamlined' to describe those smooth surface shapes (like the Rankine fairing of Section 7.3) which avoid flow separation has increasingly led to the same adjective being used also in the scientific literature. The steady flow around a streamlined shape tends to be irrotational outside a thin boundary layer.

It is interesting, for such a steady flow around a streamlined shape, to calculate the excess pressures p_e on the solid surface and to find the resultant of those excess pressures, that is to find the *drag* force with which the stream acts on the body. More accurately, this is the component of the drag force associated with normal pressures acting on the surface. However, for fluids of small viscosity like air and water, we can expect the tangential stresses due to viscosity to be quite small (Section 5.5) compared with the normal pressures and, therefore, it is of real interest to know the resultant of those normal pressures. Actually, we shall find that a quite simple argument applicable to streamlined bodies in general allows us to calculate that resultant quickly in all cases, but it is instructive to make the calculation directly at least once, as we do now for steady flow around the Rankine fairing.

The Bernoulli equation for steady flow (eqn (86) in general, or eqn (179) for irrotational flows) may be written

$$p_e + \tfrac{1}{2}\rho q^2 = \tfrac{1}{2}\rho U^2 \tag{237}$$

in a case when all streamlines emerge from a region far upstream with $q = U$ and $p_e = 0$. Applying eqn (237) with the value in eqn (236) used for the fluid speed q on the surface of the Rankine fairing, we deduce that the excess pressure on the surface is

$$p_e = \tfrac{1}{2}\rho U^2 (1 - 4s^2 a^{-2} + 3s^4 a^{-4}). \tag{238}$$

More strictly, eqn (238) represents the excess presssure in the calculated irrotational flow outside the boundary layer. It is important to note, however, that the pressure cannot vary much across a thin boundary layer. Equation (87) specifies the pressure gradient normal to a streamline as $\rho q^2/R$, where R is the streamline's radius of curvature, so that the pressure change must certainly be small compared with ρq^2 across a boundary layer which is very thin compared with the radius of curvature of the body.

For obtaining the drag (that is, the axial resultant of the excess pressures) it is convenient that eqn (238) expresses p_e in terms of the distance s from the axis. This is because the excess pressure force acting normally to a small area of the surface has a resultant in any direction (such as the axial direction) equal to the excess pressure times the *projection* of the area in that direction, as demonstrated, for example, by the geometry of Fig. 10. Therefore, the excess pressures in eqn (238) acting on the part of the body surface at distances from the axis between s and $s + ds$ must be multiplied by the axially projected area $2\pi s\,ds$ in order to obtain their resultant in the axial direction. Integrating this product from $s = 0$ to $s = a$ in order to give the total resultant, we obtain the value

$$\int_0^a \tfrac{1}{2}\rho U^2(1 - 4s^2 a^{-2} + 3s^4 a^{-4})2\pi s\,ds = \tfrac{1}{2}\pi\rho U^2(a^2 - 2a^2 + a^2) = 0. \tag{239}$$

Thus, the drag calculated from the excess pressures predicted on the assumption of steady irrotational flow is zero. Before explaining why this statement is necessarily true for all body shapes, we may briefly ask how it is that the pressure distribution in eqn (238) which is associated with the velocity distribution from eqn (236) possesses a zero resultant. The answer is that even though the excess pressure on the surface takes a maximum value $+\frac{1}{2}\rho U^2$ at the stagnation point S, it remains positive only for $s^2 < \frac{1}{3}a^2$ (that is, for one-third of the axially projected area of the body surface) and takes negative values over the rest of the surface (the other two-thirds of the projected area), with a minimum excess pressure of $-\frac{1}{8}\rho U^2$ achieved at the point M. The net effect of the positive pressures near the axis and the negative pressures ('suctions') further away from the axis evens out to zero.

Here we must comment that all results obtained for the steady flow of a uniform stream of velocity U past a stationary solid body (including, for example, the results of Sections 7.3 and 7.4 on steady flow past the Rankine fairing) may be applied without any essential change to another important problem: one of the *body* moving at constant velocity U in the opposite direction through fluid which, far from the body, is at rest. Evidently, the relative motions of body and fluid are exactly the same in these two cases which, indeed, differ only in the frame of reference being used to describe one and the same motion.

Therefore, our conclusion that the excess pressures p_e in steady flow around the Rankine fairing produce zero drag force on the body is exactly equivalent to another result: that, when the streamlined body of Fig. 47 moves to the left at constant velocity U through fluid which is at rest far from the body, *it experiences no resistance* to its motion from the action of excess pressures on the body surface. The statement in this form is furthermore readily shown to be true, on the assumption of irrotational flow, for any shape of body whatsoever by means of the following extremely general argument. (Readers may perhaps feel prepared to accept this now that they have seen the resistance calculated as zero in a case for which there are no special circumstances (e.g. of symmetry) tending to make such a conclusion appear likely.)

It is because the irrotational flow resulting from a given motion of a solid boundary is unique (Section 6.2) that the resistance associated with it must be zero. In particular, if a given body moves with constant velocity U, then at every instant the irrotational flow associated with that motion of the body is the same unique flow and possesses the same kinetic energy. (In the case of the Rankine fairing, for example, that kinetic energy may be calculated by methods given in Chapter 8 as $\frac{1}{6}\pi\rho a^3 U^2$.) If the fluid acted on the body with a resistance or drag force D then an equal force D in the opposite direction (a thrust) would be needed to move the body through the fluid against that resistance. This equal and opposite reaction of the body on the fluid would do work on the fluid at a rate DU (thrust times velocity) and the kinetic energy of the fluid would in consequence increase at this rate. Yet, if the flow remains irrotational, it retains the same unique form and the same constant kinetic energy; therefore,

$$D = 0. \tag{240}$$

This important conclusion, usually ascribed to Jean d'Alembert (1717–83), has practical relevance, of course, only to streamlined bodies, that is to bodies for which the flow remains closely irrotational because boundary layer separation is avoided. By contrast, the same argument leads to a completely different conclusion if separation occurs so as to generate vorticity in the midst of the fluid. Whilst the body moves on, then, the vortex lines remain behind (being simply convected with the fluid) and form a vortex wake of increasing length and so of increasing energy.[†] The drag is then a positive quantity D such that DU is equal to the rate of increase of the energy of that vortex wake. (Strictly, we are concerned here with *both* the kinetic energy of the vortex wake *and* any thermal energy into which some of that kinetic energy may be dissipated by viscous action.) In such a case the drag is associated with excess pressures of the general order of magnitude $\frac{1}{2}\rho U^2$ (as above) but distributed over the surface of the body in such a way that positive and negative values fail to cancel in the calculation of their resultant, so that a drag of the general order of magnitude

$$\tfrac{1}{2}\rho U^2 S, \tag{241}$$

where S is the frontal area of the body, is commonly found.

The strongly contrasting conclusion (eqn (240)) for flow around streamlined bodies is that any component of the drag associated with the distribution of excess pressures in the irrotational flow outside the thin boundary layer is necessarily zero because those pressures, although still of the general order of magnitude $\frac{1}{2}\rho U^2$, possess a zero resultant force. In practice this means that streamlined bodies experience drag of a much *smaller* order of magnitude than expression (241), such drag being exclusively associated with the action of viscosity within the boundary layer.

Often, the stark conclusion in eqn (240) from the properties of irrotational flow has been called d'Alembert's paradox because it seems to be in flat contradiction to the common experience that bodies in general are subjected to a substantial resistance when moving through a fluid. In the light of modern understanding of boundary layers, however, it should perhaps be more positively designated as d'Alembert's theorem: the first clear indication that, if 'streamlined' shapes could be designed so as to meet the stringent conditions

---

[†] The total energy of the motion is indeed equal (Section 6.3) to that of the irrotational motion which satisfies the boundary conditions plus that associated with the shed vorticity in the absence of boundary movement.

required to avoid boundary layer separation, then resistances of a very much smaller order of magnitude could be achieved.

7.5 Bluff-body flows

Pursuing a little farther the present chapter's aim to analyse various three-dimensional examples of irrotational flows, we next extend the study of fluid motion around bodies beyond the restricted case represented by a 'fairing'; that is, by a streamlined shape facing into the oncoming flow but indefinitely extended downstream. It is indeed natural to ask next how the irrotational flow around 'a finite body' (that is, a body of perfectly definite length) may differ from that around a fairing. We shall show, in fact, through examples in this section and through general theory in Chapter 8, that certain quite important differences exist.

These differences, however, do not include any enhanced tendency for either one type of flow or the other to be capable of avoiding separation: there are streamlined finite bodies just as there are streamlined fairings, while of course the vast majority of shapes chosen at random tend *not* to be streamlined (that is, the flows around them *do* exhibit separation). Proceeding now to give examples of irrotational flows about finite bodies, we could clearly continue to concentrate upon streamlined shapes, for example upon bodies of revolution which have 'front ends' (facing the oncoming flow) similar to that of Fig. 47 but which are gradually tapered at the rear so as to terminate in a sharp point or cusp. These are shapes which avoid separation, as indeed we shall verify in Chapter 9 for the corresponding shapes in two-dimensional flow.

Nevertheless, it may be more instructive at this early stage of exposition of methods for calculating irrotational flows to investigate some flows around 'bluff bodies'. These are bodies such that the steady flow around them exhibits separation, yet the calculation of an irrotational flow around a bluff body may be of some real value. For example, we know already that an impulsively started flow is *initially* irrotational; and one objective of this section is to assess whether there is any other useful application for the study of irrotational flows around finite bluff bodies.

With this objective in view, the method of Fig. 46 for constructing the Rankine fairing is given a rather obvious extension in Fig. 48, illustrating a method (also due to Rankine) for constructing an irrotational flow around a finite body: the so-called Rankine ovoid. In order that the dividing streamtube should close up again at the rear, so as to terminate that flow region occupied by the outflow from the point source which is to be replaced by a solid body, Rankine placed a point sink of equal strength J behind the point source. Then the whole outflow from the source is reabsorbed into the sink and the external flow can close up again at the rear so as to represent flow around a finite body.

With the position of the source now shifted to the point $(0, 0, -h)$, and the

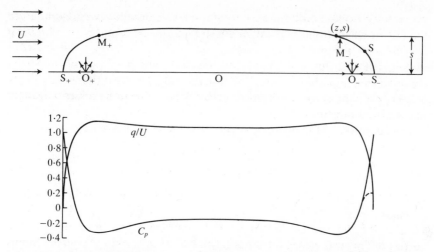

Fig. 48. Flow around the 'Rankine ovoid', specified as the dividing streamtube associated with a combination (eqn (242)) of uniform flow with a source at the point O_+, with coordinates $(0, 0, -h)$, and an equal sink at the point O_-, with coordinates $(0, 0, h)$. The flow divides at S_+ and reunites again at S_-. It is illustrated here for the case $h = 3\,(J/\pi U)^{\frac{1}{2}}$ for which the ovoid shape is almost that of two 'Rankine fairings' placed back to back. The lower diagram plots the fluid speed q on the surface in irrotational flow as a function of the axial coordinate z. This reaches maximum values of $1.16\,U$ at two points M_+ and M_-. The pressure coefficient (eqn (244)) is also plotted (i) in irrotational flow (plain line) and (ii) as modified (broken line) by boundary layer separation.

sink located at the point $(+h, 0, 0)$, the velocity potential takes the form

$$\phi = Uz - \frac{J}{4\pi[s^2 + (z+h)^2]^{\frac{1}{2}}} + \frac{J}{4\pi[s^2 + (z-h)^2]^{\frac{1}{2}}}. \tag{242}$$

This differs from the velocity potential in eqn (225) in two respects: the $r = (s^2 + z^2)^{\frac{1}{2}}$ on the right-hand side of eqn (225) is replaced by an expression for distance from the new position $(0, 0, -h)$ of the point source; while the last term on the right-hand side of eqn (242) represents the potential in eqn (212) of a point sink but with *its* location shifted to $(0, 0, +h)$.

The fact that ϕ is an *odd function* of z (in that changing the sign of z causes a simple sign change in ϕ, and therefore also in $\partial\phi/\partial s$, while leaving the axial velocity component $\partial\phi/\partial z$ unchanged) demonstrates that the flow has the perfect symmetry about the plane $z = 0$ shown in Fig. 48. In particular, the geometry around the upstream stagnation point S_+, where the axial streamline becomes a dividing streamtube, is mirrored precisely at the downstream stagnation point S_-, where the dividing streamtube reunites with the axis.

The shape of the dividing streamtube is readily found, once more, by the method of Section 7.3. Consider the flow through a surface consisting of a disc of radius s far downstream (with expression (228) representing the volume flow through it) together with the surface of a cylinder of radius s extending from the disc to where it meets the dividing streamtube at a point with cylindrical coordinates s, z (Fig. 48). This time the total volume flow through the surface in question is not J, but zero. That is because all the internal flow emerging from the point source has been reabsorbed into the sink; therefore, the entire volume flow crossing the above-mentioned surface in one direction crosses it again in the opposite direction.[†]

The equation of the dividing streamtube therefore becomes

$$0 = \pi s^2 U + \tfrac{1}{2} J \{ 1 - (z + h) [s^2 + (z + h)^2]^{-\frac{1}{2}} \}$$
$$- \tfrac{1}{2} J \{ 1 - (z - h) [s^2 + (z - h)^2]^{-\frac{1}{2}} \}. \tag{243}$$

Here, on the right-hand side, the second term represents the volume outflow through the cylindrical surface from the point source (eqn (229), but with the position of the source shifted to $(0, 0, -h)$) and the third term is the corresponding expression for the point sink at $(0, 0, h)$, with of course a minus sign attached to the latter since outflow from the sink is negative.

As in the case of eqn (231), we should add that the equation of *any* streamtube outside the dividing streamtube can be obtained from eqn (243) by replacing the left-hand side by a different constant value. In this case any *positive* constant on the left-hand side of eqn (243) will make it the equation of an external streamtube.

Figure 48 displays the shape of the Rankine ovoid (that is, the shape from eqn (243) of the dividing streamtube) in a case when the distance h of both the source and the sink from the plane of symmetry is three times as great as the distance $(J/\pi U)^{\frac{1}{2}}$ given by eqns (227) as the radius of the Rankine fairing generated by a source of strength J by itself. Actually, this is a case when each half of the Rankine ovoid has a shape very close to that of the Rankine fairing. For example, in the half with $z < 0$, the last term in eqn (243) with $h - z \geqslant 3s$ is very close to that constant value J which makes eqn (243) identical with the Rankine fairing equation (eqn (231)) except for the shift of origin to $(0, 0, -h)$.

The corresponding distribution of fluid speed q on the surface of the Rankine ovoid, calculated from the velocity potential in eqn (242) and plotted in Fig. 48, is similarly rather close to a juxtaposition of (i) the corresponding distribution (Fig. 47) for the Rankine fairing and of (ii) its mirror image in the plane of symmetry $z = 0$. Figure 48 plots also the *pressure coefficient* C_p, a convenient non-dimensional form

$$C_p = p_e / (\tfrac{1}{2} \rho U^2) \tag{244}$$

of the distribution of excess pressure (in steady flow) derived from eqn (237). This takes the value $+1$ at a stagnation point and falls to negative values in the region where the speed q exceeds U.

[†] Note that this is the case whether the cylinder meets the dividing streamtube on its right-hand half as in Fig. 48 or on its left-hand half.

It is clear from the distribution of q/U given in Fig. 48 that the conditions for irrotational flow to persist are not satisfied. At the front of the body, to be sure, the speed q rises to just a modest maximum value $1.16\,U$ at M_+ (rather as in Fig. 47), and then falls quite gradually to a value $1.06\,U$ on the plane of symmetry before rising again to the same maximum value $1.16\,U$ at M_-. Up to this point the velocity distribution has exhibited no tendency to promote boundary layer separation. Beyond M_-, however, the theoretical irrotational flow gives a velocity distribution on the surface which falls increasingly steeply to zero. Therefore, separation necessarily occurs at a point S somewhat downstream of M_-.

Wind-tunnel experiments on the flow of a uniform stream past relatively slender bodies of revolution with blunt ends like that of Fig. 48 confirm this conclusion. The pressure distribution on the surface of the model is especially easy to measure; for this purpose, small holes in the surface are attached by flexible air-filled tubes to pressure-measuring devices.

Besides confirming the presence of separation, these experiments on slender bodies of revolution demonstrate another very interesting fact, namely that the surface pressure measured at all points upstream of S is extremely close to that predicted on irrotational-flow theory. Downstream of S, on the other hand, the measured value of the pressure coefficient C_p in eqn (244) remains approximately constant (see broken line in Fig. 48), as might be expected if the external flow has separated from the surface at S leaving a relatively stagnant region or region of slow reversed flow adjacent to that part of the surface. Flow visualization confirms this interpretation. We conclude that the separation at S generates a wake of modest diameter downstream of the Rankine ovoid, and that this radically modifies the distribution of surface pressure from S onwards; but we note too that the irrotational flow upstream of S, and its associated pressure distribution, are not greatly modified by the presence of the thin wake from what they would be if the flow were everywhere irrotational.

These conclusions naturally affect the drag. The drag derived from the full-line pressure distribution associated with the theoretical irrotational flow around the Rankine ovoid must of course be zero, as the general argument given in Section 7.4 proves although symmetry makes the conclusion obvious in the present case. Therefore, it is simply the axial resultant of the pressure *defect* (amount by which the experimental broken-line curve in Fig. 48 falls below the theoretical full-line curve) that generates a drag. This is comparable with, although substantially less numerically than, the order-of-magnitude value in expression (241).[†]

All the above conclusions hold good for the case when h is relatively large (three times larger than the quantity $(J/\pi U)^{\frac{1}{2}}$) so that the Rankine ovoid is a

---

[†] As stated above, separation can be avoided on finite bodies of revolution which replace the blunt rear end of Fig. 48 by a shape tapering to a point (as used with typical airship shapes). This produces a major further reduction in drag.

slender cigar-shaped body. Smaller values of h lead to fatter ovoids (more *truly* egg-shaped!) with fatter wakes which exercise a slightly bigger influence on the upstream irrotational flow. We omit all consideration of these, however, and make now a rapid transition to a study of the extreme limiting case as h tends to zero.

At first sight, the velocity potential (eqn (242)) reverts to being a mere uniform stream $\phi = Uz$ when h tends to zero. This is not necessarily the case, however, if J is simultaneously allowed to tend to infinity. We use the concept of the *dipole* (called by some authors 'double-source' or 'doublet') which represents a combination of the source and sink of Fig. 48, and of eqn (242), in the limiting case:

$$h \to 0, \quad J \to \infty, \quad 2hJ \to G, \tag{245}$$

where G is the *strength* of the dipole. Using the idea that for a differentiable function $f(s, z)$

$$Jf(s, z+h) - Jf(s, z-h) \to G \partial f / \partial z \tag{246}$$

in this limiting case (expressions (245)), we see that the potential of a dipole is

$$\phi = -\frac{G}{4\pi} \frac{\partial}{\partial z} \frac{1}{(s^2 + z^2)^{\frac{1}{2}}} = \frac{Gz}{4\pi r^3}. \tag{247}$$

When combined, as in eqn (242), with a uniform stream of velocity U, its potential is

$$\phi = Uz + Gz/(4\pi r^3). \tag{248}$$

The irrotational flow represented by eqn (248) is of special interest because the dividing streamtube in expression (241) has a very simple form in the limiting case (expressions (245)). In fact the rule of expression (246) makes eqn (243), in that limit, become

$$0 = \pi s^2 U - \tfrac{1}{2} G \, \partial [z(s^2 + z^2)^{-\frac{1}{2}}]/\partial z$$
$$= \pi s^2 U - \tfrac{1}{2} G s^2 r^{-3}. \tag{249}$$

Discarding the solution $s = 0$ which represents the axial streamline, we obtain the equation of the dividing streamtube as the sphere

$$r = a \tag{250}$$

where, in this case, the specification of the body radius a turns out to be

$$a = (G/2\pi U)^{\frac{1}{3}}. \tag{251}$$

The velocity potential (eqn (248)) for irrotational flow around a sphere is often written in terms of a spherical polar coordinate θ (Fig. 49) such that

$$z = r \cos \theta, \quad s = r \sin \theta. \tag{252}$$

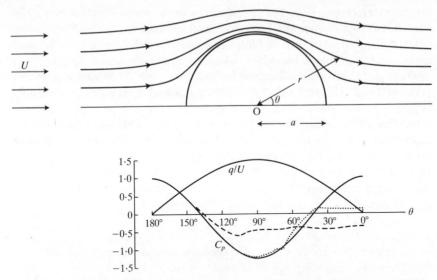

Fig. 49. Irrotational flow around a sphere, obtained as the limiting case (expressions (245)) of the flow around the Rankine ovoid. The lower diagram plots the fluid speed q on the surface as a function of the polar coordinate θ. The pressure coefficient (eqn (244)) is also plotted: (i) in irrotational flow (plain line); (ii) as modified (broken line) when the boundary layer is *laminar* at the point of separation; and (iii) as rather less modified (dotted line) at those higher Reynolds numbers for which the boundary layer becomes turbulent before separation (which, accordingly, is delayed).

Then, after the substitution given in eqn (251), the potential (eqn (248)) becomes

$$\phi = U(1 + \tfrac{1}{2}a^3 r^{-3})\, r \cos\theta. \tag{253}$$

Equation (253) makes it easy to write down the fluid speed q on the surface of the sphere as the magnitude of grad ϕ, namely

$$q = -a^{-1}(\partial\phi/\partial\theta)_{r=a} = 1.5\, U \sin\theta \tag{254}$$

(where the minus sign arises because, on the sphere, the gradient's sense is in the direction θ decreasing). By eqns (237) and (244), the corresponding theoretical pressure coefficient C_p in steady flow is

$$C_p = 1 - 2.25 \sin^2\theta. \tag{255}$$

Figure 49 shows the distributions of both q/U and C_p as a function of θ from the front stagnation point $\theta = \pi$ to the rear stagnation point $\theta = 0$.

In contrast to the cigar-shaped body of Fig. 48, we see that a much fatter bluff body such as a sphere incorporates strongly retarded flow over a *large* part of its surface. Separation occurs much farther forward, therefore, and involves a much more marked modification to the entire flow, including the irrotational part outside the boundary layer and wake.

It is in the context of such fat bluff bodies that the contrasting capabilities of laminar and turbulent boundary layers to resist separation (Section 5.6) produce particularly big differences in flow behaviour. At Reynolds numbers between about 10^3 and 10^5 there is a thin boundary layer which remains laminar up to the point of separation and which, accordingly, separates quite soon after q/U has passed its maximum value. A very wide wake results as the theoretical distribution of q/U would suggest (Fig. 49), and this greatly modifies the external irrotational flow, pushing the velocity maximum significantly upstream of $\theta = \frac{1}{2}\pi$ in such a way that separation ultimately occurs at a point close to $\theta = \frac{1}{2}\pi$ itself (as the broken-line distribution of C_p in Fig. 49 indicates). The corresponding drag coefficient

$$C_D = D/(\tfrac{1}{2}\rho U^2 S) \tag{256}$$

which relates drag to the order-of-magnitude estimate (expression (241)), is then found to have relatively *large* values (close to 0.5).

By contrast, in flow around a sphere at much higher Reynolds numbers (say, greater than 5×10^5), the boundary layer makes a transition to turbulence *before* separating from the surface. Separation is then delayed, because of the enhanced mixing effect due to turbulence (Section 5.6), until a point is reached (around $\theta = \frac{1}{4}\pi$) where a much stronger retardation of the external flow has occurred. The wake is accordingly narrower and produces considerably less modification to the external flow (as indicated by the dotted-line distribution of C_p in Fig. 49). Furthermore, the measured value of the drag coefficient C_D is close to 0.1 (or a little more) rather than to 0.5.

For flow around spheres the major change between the two flow regimes just described occurs at a so-called *critical* Reynolds number whose precise value (usually between 10^5 and 5×10^5) depends upon various factors that may tend to promote or to lessen disturbances in the boundary layer. The associated properties of the flow characteristics (including a spectacular drop in the drag coefficient C_D from values around 0.5 for sub-critical Reynolds numbers to values around 0.1 for super-critical Reynolds numbers) are extensively exploited by advanced players of ball games.

We note, finally, that the velocity potential analogous to eqn (248) or eqn (253) in a frame of reference in which the fluid far from the sphere is at rest (so that the sphere is moving through the fluid in the negative z direction) is obtained by subtracting the uniform-stream velocity potential Uz or $Ur \cos \theta$ so as to give, for example,

$$\phi = \tfrac{1}{2} U a^3 r^{-2} \cos \theta. \tag{257}$$

Of course, just as the velocity potential (eqn (253)) fails to represent accurately the steady flow past a stationary sphere, so the potential (eqn (257)) fails to represent the fluid motion associated with steady movement of the sphere through the fluid. However, this potential does have importance since the flow set up initially by impulsive movement of a body must be the irrotational flow compatible with that boundary movement (Section 6.1). Accordingly, we can say that a sphere set impulsively into motion generates initially the dipole field (eqn (257)). This is an interesting conclusion which foreshadows some of the more general results of Chapter 8 and which is itself pursued further in Section 8.3.

8
Three-dimensional far fields

We are often interested in the motions set up within a very large volume of fluid, with dimensions much greater than the maximum dimensions of any body (or bodies) whose movements may be responsible for generating the fluid motions. It is frequently convenient to model such a system as an expanse of fluid which is externally unbounded (that is, extending indefinitely in all directions) and which is disturbed only as a result of the movements of an object (or objects) of limited size situated within it. Such a model assumes that the fluid velocity tends to zero in the *far field*, that is at distances large compared with the size of the agency causing motion.

It turns out (Sections 8.1 and 8.2) that irrotational far fields are necessarily of one or other of two distinct types. The first type arises when and only when the *volume* of an object (or objects) whose movement generates the fluid motion is changing. At large distances from such an object, the flow field becomes more and more closely that of a *sink* if the object, as with the cavity of Section 7.2, is contracting, or that of a *source* if the object is expanding. In other cases, including the important case when a solid body of unchanging volume moves in a fluid, an irrotational far field is necessarily of the second type; i.e. one which at large distances becomes more and more closely like that of the *dipole* defined in Section 7.5. In general, of course, it is *only* the far fields that have these simple forms; thus, although a spherical cavity contraction generates everywhere (Section 7.2) a fluid motion of sink type, and an impulsively moving rigid sphere generates everywhere (Section 7.5) a fluid motion of dipole type, similar actions of non-spherical bodies produce more complicated motions. Yet in the far field, the difference from a simple source or sink motion (or a dipole motion if the body volume is not changing) becomes a vanishingly small proportion of the whole velocity field as distance from the body increases.

The above results are first used to extend all previous conclusions (Chapter 6) regarding uniqueness of flows to the case of unbounded fluids whose motion vanishes at large distances. Then, for irrotational motions generated by the uniform motion of a rigid body, some useful and interesting relations are derived (Sections 8.3 and 8.4) between overall flow properties and the strengths of far-field dipoles. Finally (Section 8.5), the far-field properties of steady flows past stationary bodies that leave wakes are investigated and shown to involve reemergence, in the irrotational flow outside the wake, of a far field of source type, while methods for the use of far-field results to estimate wind-tunnel interference and other perturbing effects of remote boundaries are briefly noted.

8.1 Spherical means and Green's formula

The analysis establishing the main results indicated above is valuable in its own right because it uses the fruitful concept of spherical means and because it derives Green's formula, which is an important asset for the computation of irrotational flows. There is also a powerful mathematical interest in arguments showing that, if an irrotational velocity field tends to zero at large distances, then it *must* do so at least as fast as the inverse square of the distance.

The Divergence Theorem (eqn (48)), applied to any region of fluid L where the velocity field takes the irrotational form $\mathbf{u} = \operatorname{grad} \phi$, tells us (since Laplace's equation (eqn (207)) makes the left-hand side zero) that

$$\int_{\partial L} (\partial\phi/\partial n)\,\mathrm{d}S = 0. \tag{258}$$

This result, used for several different regions L in the present section, gives a specially interesting conclusion when L is the interior of a *spherical* surface ∂L, of centre P and radius r, which we shall call Σ_r.

The average, or mean, value of the potential over the spherical surface Σ_r (of area $4\pi r^2$) is called the *spherical mean*

$$\bar{\phi} = (4\pi r^2)^{-1} \int_{\Sigma_r} \phi\,\mathrm{d}S. \tag{259}$$

Here, an element of area $\mathrm{d}S$ on the spherical surface Σ_r of radius r can always be written

$$\mathrm{d}S = r^2\,\mathrm{d}\Omega, \tag{260}$$

where $\mathrm{d}\Omega$ is the element of solid angle subtended at the centre of the sphere by that element of area. In eqn (260), any readers unfamiliar with the concept of solid angle can simply regard $\mathrm{d}\Omega$ as a shorthand for $\sin\theta\mathrm{d}\theta\mathrm{d}\psi$, where spherical polar coordinates (r, θ, ψ) centred upon P are used to specify position so that $r^2 \sin\theta\,\mathrm{d}\theta\,\mathrm{d}\psi$ represents an element of area of Σ_r. On either basis, the spherical mean (eqn (259)) becomes

$$\bar{\phi} = (4\pi)^{-1} \int_{\Sigma_r} \phi\mathrm{d}\Omega, \tag{261}$$

an integral over the total solid angle $4\pi = \int \mathrm{d}\Omega$ subtended at P.

Since $\mathrm{d}\Omega$ does not depend on r, we can calculate the rate of change of $\bar{\phi}$ with respect to r as

$$\mathrm{d}\bar{\phi}/\mathrm{d}r = (4\pi)^{-1} \int_{\Sigma_r} (\partial\phi/\partial r)\,\mathrm{d}\Omega, \tag{262}$$

where the partial derivative $\partial\phi/\partial r$ is taken for a fixed element $\mathrm{d}\Omega$ of solid angle or, in spherical polar coordinates, for fixed values of θ and ψ. This makes it identical with $\partial\phi/\partial n$ on Σ_r, so that eqn (258) can be used to show that $d\bar{\phi}/dr$ vanishes; thus by eqns (260) and (262)

$$d\bar{\phi}/dr = (4\pi r^2)^{-1} \int_{\Sigma_r} (\partial\phi/\partial n)\,dS = 0. \tag{263}$$

In words, the mean of ϕ over *any spherical surface centred on* P *inside which Laplace's equation is satisfied* is the same. In particular, it must be the same as the limit of the mean over a vanishingly small sphere Σ_r centred on P (the limit as $r \to 0$, which is evidently the value ϕ_P of ϕ at P itself), so we have proved the important property

$$\bar{\phi} = \phi_P \tag{264}$$

of solutions of Laplace's equation, that the spherical mean is equal to the value at the centre.

Before using this result directly, we may note two interesting consequences of applying it to *derivatives* of ϕ (actually, $\partial\phi/\partial x$ and $\partial\phi/\partial t$) which, of course, also satisfy Laplace's equation. We show first that, in irrotational flow, the fluid speed q_P at a point P is always *less than or equal to* the average fluid speed $\bar{q}$ over the surface Σ_r of a fluid sphere centred on P. This demonstration is immediate if we make a special choice of axes with the x axis in the direction of the fluid velocity at P. Then

$$q_P = (\partial\phi/\partial x)_P = \overline{\partial\phi/\partial x} \leqslant \bar{q}, \tag{265}$$

since on the surface Σ_r we necessarily have

$$q = [(\partial\phi/\partial x)^2 + (\partial\phi/\partial y)^2 + (\partial\phi/\partial z)^2]^{\frac{1}{2}} \geqslant \partial\phi/\partial x. \tag{266}$$

Secondly, we demonstrate that in irrotational flow the pressure p_P at P is always *greater than or equal to* the spherical mean $\bar{p}$ of the pressure over the same surface Σ_r. Once again, this demonstration is immediate if we make a special choice of axes, this time choosing a set of moving axes in which the fluid speed q_P at P is zero. Equation (178), with the excess pressure p_e substituted from eqn (70), gives the pressure p in the form

$$
\begin{aligned}
p &= p_0 - \rho g H + p_e \\
&= p_0 - \rho g H + \rho f(t) - \rho \partial\phi/\partial t - \tfrac{1}{2}\rho q^2,
\end{aligned} \tag{267}
$$

so that in the chosen axes (with $q_P = 0$) we have

$$p_P = p_0 - \rho g H_P + \rho f(t) - \rho(\partial\phi/\partial t)_P. \tag{268}$$

On the other hand, the spherical mean $\bar{p}$ is

$$\bar{p} = p_0 - \rho g \bar{H} + \rho f(t) - \rho\overline{\partial\phi/\partial t} - \tfrac{1}{2}\rho\overline{q^2} \tag{269}$$

and this is less than or equal to

$$p_0 - \rho g \bar{H} + \rho f(t) - \rho\overline{\partial\phi/\partial t}, \tag{270}$$

which in turn is equal to p_P since the spherical mean of a solution $\partial\phi/\partial t$ of Laplace's equation is equal to its value at the centre (and the same is evidently true also of the height H).

The above results $q_P \leqslant \bar{q}$ and $p_P \geqslant \bar{p}$ for irrotational flow mean that any point P where the fluid speed q is a *local maximum* must lie on the *boundary*; similarly any *local minimum* of the pressure p must be found on the boundary (not necessarily at the same point). For if P were in the interior of the fluid then a small fluid sphere with centre P and surface Σ_r could be found with $q_P > \bar{q}$ or $p_P < \bar{p}$ (the local maximum or minimum property, respectively) in contradiction to our general results for irrotational flows.[†]

We next use the ideas of spherical means in an extended form and in association with other important ideas originated by George Green (1793–1841), in order to prove the validity of *Green's formula* for the velocity potential in externally unbounded fluid under the proviso that the fluid velocity tends to zero at large distances from those internal boundaries whose actions generate the irrotational flow. Then, alongside brief references to wider uses of Green's formula, we concentrate on applying it (Section 8.2) to deriving the properties of irrotational far fields.

The approach introduced by Green utilizes the Divergence Theorem (eqn (48)) applied, rather as in Section 6.2, to the velocity vector $\mathbf{u} = \operatorname{grad} \phi$ *multiplied by a velocity potential*; but multiplied this time by the potential Φ of a *different* velocity field. Since

$$\operatorname{div}(\Phi\mathbf{u}) = \Phi \operatorname{div}\mathbf{u} + (\operatorname{grad}\Phi)\cdot\mathbf{u} = (\operatorname{grad}\Phi)\cdot(\operatorname{grad}\phi), \qquad (271)$$

where (as in eqn (188)) the equation of continuity has been used, the Divergence Theorem (eqn (48)) becomes

$$\int_L (\operatorname{grad}\Phi)\cdot(\operatorname{grad}\phi)\,dV = \int_{\partial L} \Phi\mathbf{u}\cdot\mathbf{n}\,dS = \int_{\partial L} \Phi(\partial\phi/\partial n)\,dS. \qquad (272)$$

Now, the left-hand side of eqn (272) is unaltered if we interchange ϕ and Φ, which gives us the by no means obvious result that the right-hand side has the same property. This result, often known as Green's theorem, states that

$$\int_{\partial L} (\Phi\partial\phi/\partial n - \phi\partial\Phi/\partial n)\,dS = 0, \qquad (273)$$

if both potentials ϕ and Φ satisfy Laplace's equation (the equation of continuity for each irrotational velocity field) throughout the region L.

We now consider an irrotational motion of fluid which is externally

---

[†] Actual flows, of course, can at most be irrotational outside boundary layers, across which the fluid speed q changes steeply although the pressure p changes very little. Accordingly, the conclusion that any local minimum of p must be found on the boundary remains closely correct for the flow as a whole in such cases, but the corresponding result for q applies *only* to the flow outside the boundary layer.

unbounded and in which the fluid is disturbed entirely as a result of actions taking place within an internal boundary S of limited size. Here, for example, S may be the surface of a solid body; or it may be a surface separating the irrotational flow outside it from motions with non-zero vorticity that are present inside S. Given that the velocity field is irrotational outside S and that its magnitude tends to zero as the distance from S increases indefinitely, we seek an expression for ϕ_P, the velocity potential at any point P inside the irrotational flow, in the form of a surface integral over S.

With this objective, Green's theorem (eqn (273)) is applied with Φ taken as the potential

$$\Phi = r^{-1} \tag{274}$$

of a sink of strength 4π situated at P. This satisfies Laplace's equation throughout any region which excludes P itself. Here, because P is *inside* the irrotational flow, we can choose a small sphere Σ_ε of centre P and radius ε which lies entirely within that flow and apply eqn (273) to a region L bounded internally by *both* S and Σ_ε (so that *both* ϕ and Φ satisfy Laplace's equation throughout L). Since L must furthermore be a finite region if the Divergence Theorem (leading to eqn (273)) is to be applied within it, we take L as bounded externally by a very large spherical surface Σ_R with centre P and radius R (Fig. 50). Thus, the boundary ∂L consists of (i) an external boundary Σ_R with the outward normal $\mathbf{n}$ pointing away from P; (ii) an internal boundary Σ_ε on which the outward normal from L points *towards* P; and (iii) another internal boundary S on which the outward normal from L is again a vector $\mathbf{n}$ pointing into the *interior* of S (Fig. 50). Thus eqn (273) can be written

$$\left(\int_{\Sigma_R} + \int_{\Sigma_\varepsilon} + \int_S \right) [r^{-1} \partial\phi/\partial n - \phi\partial(r^{-1})/\partial n] \, dS = 0. \tag{275}$$

Equation (275) must be accurate whatever the value of ε (provided that the interior of Σ_ε lies outside S) and whatever the value of R (provided that the interior of Σ_R includes the whole of S). This shows clearly that the integral over Σ_ε must actually be independent of ε, and the integral over Σ_R independent of R, when the above conditions are satisfied. Nevertheless, we shall verify these statements directly by using the concepts of spherical means.

The integral over Σ_ε in eqn (275), with the normal pointing *towards* P (along the direction r decreasing) so that $\partial/\partial n = -\partial/\partial r$, can be written

$$\varepsilon^{-1} \int_{\Sigma_\varepsilon} (\partial\phi/\partial n) \, dS + \int_{\Sigma_\varepsilon} [\phi\partial(r^{-1})/\partial r] \, dS$$

$$= 0 - \varepsilon^{-2} \int_{\Sigma_\varepsilon} \phi \, dS = -4\pi\bar{\phi} = -4\pi\phi_P, \tag{276}$$

using eqn (258) applied to the interior of Σ_ε, as well as eqns (259) and (264), to

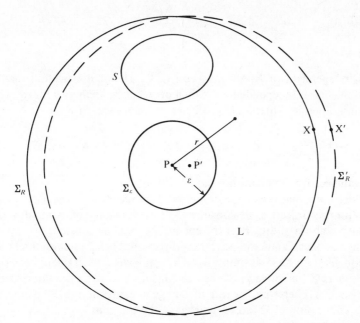

Fig. 50. Green's formula is obtained by applying Green's theorem (eqn (273)) to a region L bounded *externally* by a sphere Σ_R of centre P and large radius R, and *internally* by a surface S and by a sphere Σ_ε of centre P and small radius ε. The integrals over Σ_ε and Σ_R are first shown to be independent of the actual radii ε and R. Then the integral over Σ_R is also shown to be independent of the position of the centre P by considering the effect of moving that centre to a nearby position P'.

show that the expression is indeed independent of the value of ε. However, we cannot use the same analysis to obtain the integral over Σ_R because Laplace's equation is satisfied by ϕ only in a *part* of the interior of Σ_R (the part outside S).

The integral over Σ_R in eqn (275), with the normal *pointing away from* P so that $\partial/\partial n = \partial/\partial r$, can be written

$$R^{-1}\int_{\Sigma_R}(\partial\phi/\partial n)\,\mathrm{d}S - \int_{\Sigma_R}[\phi\partial(r^{-1})/\partial r]\,\mathrm{d}S = JR^{-1} + R^{-2}\int_{\Sigma_R}\phi\,\mathrm{d}S, \qquad (277)$$

where J is the *rate of volume outflow* produced by actions with S. Here it is evident that J must be independent of R, since the Divergence Theorem (eqn (258)) gives

$$J = \int_{\Sigma_R}(\partial\phi/\partial n)\,\mathrm{d}S = -\int_S(\partial\phi/\partial n)\,\mathrm{d}S. \qquad (278)$$

Thus, the integral over Σ_R as given by eqn (277) can be written, using eqn (260), as

$$JR^{-1} + \int_{\Sigma_R} \phi\,\mathrm{d}\Omega \tag{279}$$

with J (independent of R) given by eqn (278). The demonstration that this expression (279) is independent of R then proceeds by analogy with eqn (263), by writing down its differential coefficient with respect to R as

$$-JR^{-2} + \int_{\Sigma_R} (\partial\phi/\partial r)\,\mathrm{d}\Omega = -JR^{-2} + R^{-2}\int_{\Sigma_R}(\partial\phi/\partial n)\,\mathrm{d}S, \tag{280}$$

which vanishes by the definition (eqn (278)) of J.

We next show that the expression (279) *is independent of the position of* P *itself*. With this objective, we consider the rate of change of expression (279) when, although the radius R of the sphere Σ_R remains constant, its centre P moves just a short distance in (say) the x direction. Then J remains unchanged (since eqn (278) expresses it simply as an integral over S itself) while each point on the sphere Σ_R is moved just the same distance in the x direction as is its centre (Fig. 50). Therefore, the rate of change of expression (279) with respect to the x coordinate of P can be written as the integral

$$\int_{\Sigma_R}(\partial\phi/\partial x)\,\mathrm{d}\Omega = 4\pi\,\overline{\partial\phi/\partial x}, \tag{281}$$

of $\partial\phi/\partial x$ over the total solid angle 4π subtended at P. However, the value of expression (279) is independent of R, so we can take R as large as we please in evaluating eqn (281). Finally, because the velocity component $\partial\phi/\partial x$ tends to zero as R becomes indefinitely large, its spherical mean must also tend to zero. Therefore, the only possible value of eqn (281), which is known to be independent of R and can be shown to be as small as we like by taking R large enough, is zero.

This completes the proof that, in the form (275) of Green's theorem, the integral over Σ_R (given by expression (279)) is a mere constant independent of the position of P. Writing that constant as $4\pi C$ and using the eqn (276) for the integral over Σ_ε, we can finally rewrite eqn (275) (after division by 4π) as Green's formula

$$\phi_P = C + (4\pi)^{-1}\int_S [r^{-1}\partial\phi/\partial n - \phi\,\partial(r^{-1})/\partial n]\,\mathrm{d}S, \tag{282}$$

where the normal $\mathbf{n}$ points into the interior of S (Fig. 50) and where we use r throughout this section to signify *distance from the point* P. Here, the presence of an unknown constant C in the velocity potential is not unexpected since for a given velocity field grad ϕ there is always an arbitrary constant present in the potential ϕ.

Green's formula (eqn (282)) has proved valuable as a foundation for many methods of computing irrotational flows which proceed by reducing the problem to the solution of an integral equation. For example, when a boundary condition prescribes the value of $\partial\phi/\partial n$ on S, one may consider using the limit of eqn (282) as P tends to a point on S itself as an integral equation to determine the values of ϕ on S, after which the general form of eqn (282) gives the value of ϕ at a general point P. In this chapter, however, we use eqn (282) simply as a means of studying irrotational far fields.

8.2 Source and dipole far fields

A major change of notation is made at this point of the chapter. Whereas in Section 8.1 (with its spherical means) everything was centred on P, so that it was convenient to use the simple symbol r to mean 'distance from P' at all times, we now deliberately select a new centre for our analysis of the irrotational flow. We choose rather to centre that analysis *on an origin of coordinates* O situated somewhere inside S (Fig. 51). We use the letter Q to signify a typical point on S, and we rewrite Green's formula (eqn (282)) with the symbol r_{PQ} (distance from P to Q) replacing r, so as to give

$$\phi_P = C + (4\pi)^{-1} \int_S [r_{PQ}^{-1} \partial\phi/\partial n - \phi\partial(r_{PQ}^{-1})/\partial n]\,dS. \tag{283}$$

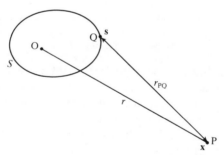

Fig. 51. Illustrating the notation used to obtain the far-field behaviour of Green's formula (283), where a typical point Q on S has position vector s relative to the origin O, while a point P in the far field has position vector x and distance r from O. The distance PQ is written $r_{PQ} = |\mathbf{x} - \mathbf{s}|$.

By contrast, the letter r is henceforth used to denote the distance of P from the origin O. In fact the vector x, with magnitude r, is taken as the position vector of P relative to O, and the position vector of Q (a point on the surface S) relative to O is taken as s. This makes

$$r_{PQ} = |\mathbf{x} - \mathbf{s}|, \tag{284}$$

the magnitude of the displacement $\mathbf{x} - \mathbf{s}$ from Q to P.

A far-field analysis of eqn (283) can make use of the fact that s remains of limited magnitude (limited by the maximum distance of a point on S from the chosen origin O within it), while the magnitude r of x increases without limit in the far field. Thus, s is small in magnitude compared with x so that r_{PQ}^{-1} differs only slightly from r^{-1} (where r is now the magnitude of x).

More accurately, we can estimate the difference between r_{PQ}^{-1} and r^{-1} when P is in the far field from the fact that the gradient of r^{-1} is

$$\nabla(r^{-1}) = -xr^{-3}, \tag{285}$$

where we note that the quantity in eqn (285) is $O(r^{-2})$, a symbolism meaning that it is 'of order at most r^{-2} as r increases indefinitely'. When s is of small magnitude compared with x, we can approximate r_{PQ}^{-1} in terms of the gradient in eqn (285) as

$$r_{PQ}^{-1} = |x-s|^{-1} = r^{-1} - s \cdot \nabla(r^{-1}) + O(r^{-3})$$
$$= r^{-1} + (s \cdot x)r^{-3} + O(r^{-3}), \tag{286}$$

where the error term is here $O(r^{-3})$ because the next term in a Taylor expansion of $|x-s|^{-1}$ with respect to s would involve *second* derivatives of r^{-1}.

The eqn (285) for the gradient may also be used to simplify the normal derivative of r_{PQ}^{-1} in eqn (283), since this is the normal component of the gradient with respect to s (which we write ∇_s) of r_{PQ}^{-1},

$$\partial(r_{PQ}^{-1})/\partial n = n \cdot \nabla_s |x-s|^{-1} = n \cdot (x-s)|x-s|^{-3}$$
$$= (n \cdot x)r^{-3} + O(r^{-3}). \tag{287}$$

Here we see that the *leading* term is $O(r^{-2})$ (in contrast to eqn (286) with its r^{-1} leading term) and that the error in replacing $x-s$ by x in that leading term, when the magnitude of s is small compared with that of x, is of the order $O(r^{-3})$ (which we are already neglecting).

Using eqns (286) and (287) to find a far-field approximation to Green's formula (eqn (283)) with error $O(r^{-3})$, we obtain

$$\phi_P = C + (4\pi r)^{-1} \int_S (\partial\phi/\partial n)\,dS + (4\pi r^3)^{-1} x \cdot \int_S s(\partial\phi/\partial n)\,dS$$

$$- (4\pi r^3)^{-1} x \cdot \int_S n\phi\,dS + O(r^{-3}). \tag{288}$$

Now the first integral in eqn (288), with n pointing into the interior of S, has already been defined in eqn (278) as $-J$, where J signifies the rate of volume outflow produced by actions within S. The last integral and the last but one in eqn (288) may be combined (since they have the same multiplier outside) to give a vector

$$G = \int_S (n\phi - s\partial\phi/\partial n)\,dS, \tag{289}$$

whose significance has yet to be determined, and the result can then be written

$$\phi_P = C - J/(4\pi r) - G \cdot x/(4\pi r^3) + O(r^{-3}). \tag{290}$$

Equation (290) identifies the constant C as simply the limiting value of ϕ_P as $r \to \infty$. This reinforces the view of C as just an arbitrary constant that is necessarily present in a velocity potential ϕ defined by the property grad ϕ = **u**. We see now that it must always be possible to fix that constant C as zero by requiring a choice of ϕ which, besides satisfying grad ϕ = **u**, possesses a zero limiting value as $r \to \infty$. Whether or not this choice is made, the value of C cannot affect in any way the fluid velocity **u** = grad ϕ in the far field.

In fact, that velocity in the far field, which is assumed in Section 8.1 to tend to zero, is seen from eqn (290) to tend to zero at least as fast as $O(r^{-2})$ if $J \neq 0$, and at the still faster rate $O(r^{-3})$ if $J = 0$. These cases represent the two types of far field mentioned at the beginning of this chapter.

The first type arises when there is a non-zero (positive or negative) rate of volume outflow J produced by motions within S. Then the far field is dominated by the term

$$-J/(4\pi r) \tag{291}$$

in eqn (290), which represents the potential (eqn (215)) of a point source of strength J situated at the origin. (The source becomes a sink, of course, if J is negative.) Thus, a rate of volume outflow J, even if distributed over an *extended* area S, must 'look like' a point source of strength J from far enough away (always provided, of course, that the flow is irrotational).

The second type of far field arises in the quite common case in which there is no net volume outflow from within S (in fact for an incompressible fluid such a net outflow is possible *only* if one or more foreign objects within S are changing their volume, like the cavity of Section 7.2). In this case $J = 0$, the third term in eqn (290) dominates, and we see that it represents a slightly generalised form of the *dipole* field in eqn (247).

Thus, whereas in Section 7.5 the closely neighbouring sink and source which in the limiting cases (245) became the dipole in eqn (247) were separated by a displacement $2h$ in the negative z direction, that displacement for a general dipole is given in the *vector* form $2\mathbf{h}$, whose direction is arbitrary while its magnitude undergoes a similar limiting process. With the source of strength J at the point $\mathbf{x} = +\mathbf{h}$ and the sink of equal strength at $\mathbf{x} = -\mathbf{h}$ (Fig. 52—note that the case discussed in Section 7.5 would have $\mathbf{h} = (0, 0, -h)$), we consider the limiting case

$$\mathbf{h} \to 0, \quad J \to \infty, \quad 2\mathbf{h}J \to \mathbf{G}. \tag{292}$$

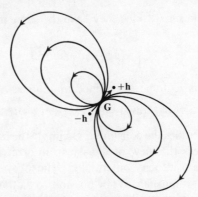

Fig. 52. The dipole field obtained from a source of strength J at the point $\mathbf{x} = +\mathbf{h}$ and a sink of equal strength at $\mathbf{x} = -\mathbf{h}$, after the limiting process (expressions (292)) has been applied.

On this basis, the dipole strength $\mathbf{G}$ is a *vector* whose magnitude has the same meaning as before and which points in the direction from the sink towards the source.

For any differentiable function $f(\mathbf{x})$ the limiting process (292) gives the result

$$Jf(\mathbf{x}+\mathbf{h}) - Jf(\mathbf{x}-\mathbf{h}) \to \mathbf{G}\cdot\nabla f \tag{293}$$

so that the potential ϕ of a dipole of strength $\mathbf{G}$ takes the form

$$\phi = \mathbf{G}\cdot\nabla(1/4\pi r) = -\mathbf{G}\cdot\mathbf{x}/(4\pi r^3). \tag{294}$$

The velocity potential in eqn (294) gives instantaneous streamlines as plotted in Fig. 52. They are, of course, identical with the lines of force of a magnetic dipole.

In short, the *second* type of irrotational far field (given by eqn (290) with $J = 0$) takes the form of a dipole field with the strength $\mathbf{G}$ given by the integral in eqn (289). Here we postpone any discussion of the physical interpretation of that integral to later sections, but we may note that, in the most interesting cases to be studied there, $\mathbf{G}$ takes a non-zero value so that it does not become necessary to proceed to any higher-order term in the far-field approximation.

In the meantime, we may conclude this section by extending all the *uniqueness* results proved in Chapter 6 for finite regions of fluid to the case of externally unbounded fluid with the fluid velocity tending to zero at large distances. Actually, all of the uniqueness results in Chapter 6 were derived very simply and directly from one such result, namely that if the normal velocity $\mathbf{u}\cdot\mathbf{n}$ is zero on the boundary of the fluid region then the velocity is zero throughout the region, and it will suffice if we extend the proof (eqns (186) to

(190)) of this result to the case of unbounded fluid with the velocity tending to zero at large distances.

We do this once again by applying the Divergence Theorem (eqn (48)) to the vector $\phi\mathbf{u}$ in a finite region L, but this time we take L as bounded externally by a fluid sphere Σ_R of very large radius R *and centre the origin* as well as internally by the actual physical boundary S of the fluid. The boundary ∂L then consists of a part S on which $(\phi\mathbf{u})\cdot\mathbf{n} = 0$, as in eqn (186), and another part for which we can use the far-field description (eqn (290)) with $J = 0$ (since $\mathbf{u}\cdot\mathbf{n} = 0$ on S). The Divergence Theorem therefore gives

$$\int_L (\mathbf{u}\cdot\mathbf{u})\,\mathrm{d}V = \int_L \mathrm{div}\,(\phi\mathbf{u})\,\mathrm{d}V = \int_{\Sigma_R} \phi\,(\mathbf{u}\cdot\mathbf{n})\,\mathrm{d}S, \tag{295}$$

where the integral is over a sphere of area $4\pi R^2$, but the integrand decreases at least as rapidly as R^{-3} (since $\mathbf{u}$ is a far-field velocity of dipole type while ϕ tends to the constant value C). Therefore the eqn (295) has a right-hand side which is $O(R^{-1})$ and can be shown to be as small as we please by taking R large enough; and so the left-hand side must be zero, from which we deduce as before that $\mathbf{u}$ must be zero everywhere.

From this conclusion for externally unbounded fluid, all of the uniqueness results of Chapter 6 can be immediately extended to that case. Given, for example, two velocity fields $\mathbf{u}_1$ and $\mathbf{u}_2$ which have the same vorticity distribution $\boldsymbol{\omega}$ in the region outside the solid boundary S and the same motion $\mathbf{u}_s$ of that solid boundary, eqns (191) and (192) once more apply so that the difference $\mathbf{u} = \mathbf{u}_1 - \mathbf{u}_2$ satisfies the eqns (193) which have just been shown, even in externally unbounded fluid, to make $\mathbf{u}$ zero everywhere. Therefore in this case, too, the velocity field $\mathbf{u}_1$ and $\mathbf{u}_2$ must simply be the same; in short, such conditions specify the flow uniquely.

8.3 Energy, impulse, and added mass

We concentrate now on the physical interpretation of the dipole strength $\mathbf{G}$ in the case $J = 0$ when it dominates the eqn (290) for a flow's far field. We relate it to important quantities such as the kinetic energy of the flow field, the impulsive force (or 'impulse') required to set up that flow field from rest, and the effective 'added mass' imparted by the presence of the fluid to a body immersed within it.

Equation (289) expresses $\mathbf{G}$ in two parts:

$$\mathbf{G} = \mathbf{G}_1 + \mathbf{G}_2, \text{ where } \mathbf{G}_1 = \int_S \mathbf{n}\phi\,\mathrm{d}S \quad \text{and} \quad \mathbf{G}_2 = -\int_S \mathbf{s}(\partial\phi/\partial n)\,\mathrm{d}S \tag{296}$$

have different physical interpretations. The second integral $\mathbf{G}_2$ is related to the *centroid of the displaced fluid.* This phrase refers to the fact that fluid in irrotational motion just fills the space outside the surface S, so that we can imagine the region D inside S (whether it contains a solid body, or fluid carrying vorticity, or both) as a region from which irrotationally moving fluid has been displaced. At the boundary S of this region D of displaced fluid the component of irrotational motion normal to the boundary is $\partial\phi/\partial n$; which moves that surface S, separating the irrotationally moving fluid from the region D, at a velocity $\partial\phi/\partial n$ normal to itself (Fig. 53). This in turn moves D, and may have the effect of shifting the centroid of D.

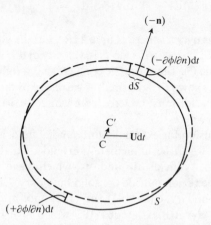

Fig. 53. This illustrates the motion of the centroid C of the displaced fluid. While fluid motions shift S from its position (plain line) at time t to a new position (broken line) at time $t + dt$, the centroid of the fluid supposed to have been displaced by S moves from C to C'.

The position vector $\mathbf{x}_c$ of the centroid of the region D of displaced fluid is given by the equation

$$V\mathbf{x}_c = \int_D \mathbf{x}\, dV, \qquad (297)$$

where V is the volume of D. Here we should perhaps note that, when D is occupied by a solid body, that body's density need not be uniformly distributed, so that the centroid $\mathbf{x}_c$ is not necessarily the body's centre of mass. Nevertheless, the fluid which is presumed to have been displaced by D can be thought of having the same *uniform* density as the fluid outside D, so we may properly describe $\mathbf{x}_c$ as the centre of mass of the displaced fluid.

We now show that the rate of change of eqn (297) with respect to time is $\mathbf{G}_2$. In fact, in a small time dt, while the main part of the volume D remains unchanged, a small area of surface dS moving outwards with normal velocity $-\partial\phi/\partial n$ (where the minus sign results because in eqns (296) the normal $\mathbf{n}$ is taken as directed *inwards* into D) produces a local *increase* in that volume (Fig. 53) by an amount

$$dS(-\partial\phi/\partial n)\,dt, \tag{298}$$

an increase, that is, for $\partial\phi/\partial n$ negative which, naturally, becomes a decrease where $\partial\phi/\partial n$ is positive. The total volume V remains unchanged since

$$J = \int_S (-\partial\phi/\partial n)\,dS = 0. \tag{299}$$

From expression (298) we see that, if the position vector $\mathbf{x}$ takes the value $\mathbf{s}$ at a point on S, the integral (eqn (297)) is increased by an amount

$$\left[\int_S \mathbf{s}(-\partial\phi/\partial n)\,dS\right]dt = \mathbf{G}_2\,dt \tag{300}$$

in time dt. Thus, the *rate* of change of eqn (297) is

$$\mathbf{G}_2 = V\,d\mathbf{x}_c/dt = V\mathbf{U}, \tag{301}$$

where $\mathbf{U}$ is the velocity of the centroid of the displaced fluid.

Similarly, the first integral $\mathbf{G}_1$ in eqns (296) has a simple physical interpretation in terms of the total impulse $\mathbf{I}$ which has to be imparted to set up the irrotational flow instantaneously from rest. We know that the motion outside S is the unique irrotational motion compatible with the instantaneous distribution of normal velocity $\partial\phi/\partial n$ on S, and we know also (Section 6.1) that, if the fluid were initially at rest, then impulsive actions to set up this distribution of boundary velocities would instantaneously generate that unique irrotational flow. We can calculate the total impulse $\mathbf{I}$ exerted on the fluid during this process from eqn (170) which gives $\mathbf{I}$ as the integral of the force exerted $\mathbf{F}$ over the very small time τ of its action. This force, of course, must be exerted through excess pressures p_e acting at the surface S on the fluid outside it. On an element dS of the surface the force acting is $p_e\,dS$ directed into the fluid along the outward normal $(-\mathbf{n})$, and the corresponding impulse (time integral of this force) is

$$\left(\int_0^\tau p_e\,dt\right)(-\mathbf{n})\,dS. \tag{302}$$

Equation (173), however, allows us to rewrite this as

$$\rho\phi\mathbf{n}\,dS \tag{303}$$

in terms of the velocity potential of the irrotational flow, and hence to express the total impulse $\mathbf{I}$ acting on the fluid (namely, the integral of expression (303) over S) as

$$\mathbf{I} = \rho\int_S \mathbf{n}\phi\,dS = \rho\mathbf{G}_1. \tag{304}$$

Equations (296), (301), and (304) may now be combined to produce a most important conclusion, as follows. *In irrotational flows with dipole far fields, the dipole strength* **G** *is given by the equation*

$$\rho \mathbf{G} = \mathbf{I} + \rho V \mathbf{U} \tag{305}$$

expressing $\rho\mathbf{G}$ as the sum of (i) the impulse **I** required to set up the irrotational flow instantaneously from rest, and (ii) the momentum of the hypothetical 'displaced fluid' (that is of fluid with uniform density ρ imagined to be filling the region D not occupied by the irrotational flow; a region of volume V whose centroid moves with velocity **U**).

In this striking result it is undoubtedly tempting to make a direct analogy between the elements (i) and (ii) in the expression for $\rho\mathbf{G}$, on the argument that the impulse **I** must surely generate a total momentum **I** in the irrotational flow. Actually, however, for externally unbounded fluid the concept of its 'total momentum' is not a well defined quantity. For example, the volume integral of $\rho\nabla\phi$ over an externally unbounded region of fluid is not a convergent integral for a dipole far field with $\nabla\phi$ of order r^{-3} as $r \to \infty$. In all cases however, the *impulse* **I** required to set up the flow is well defined, and its direct relationship to the strength of the dipole far field is valuable and interesting.

Furthermore, the impulse **I** does possess one of the most useful properties characteristic of 'momentum'. This is that when an irrotational flow field is changing continuously, because the motions of the boundary S are changing, the *force* **F** with which the boundary S acts on the fluid satisfies an equation

$$d\mathbf{I}/dt = \mathbf{F} \tag{306}$$

in the form of Newton's second law. The change in **I** in a very small time dt can indeed by eqn (170) be written as

$$d\mathbf{I} = \int_0^{\tau + dt} \mathbf{F}\, dt - \int_0^{\tau} \mathbf{F}\, dt = \mathbf{F}\, dt, \tag{307}$$

where the second integral represents the impulse **I** due to the forces **F** necessary to set up an irrotational flow in the very short time τ, and the first integral represents the impulse $\mathbf{I} + d\mathbf{I}$ needed to generate that irrotational flow first of all and then go on through the action of the required force **F** over a further very short time dt to generate the very slightly changed irrotational flow. Equation (306) means that the concept of flow impulse is really valuable for calculating the forces acting between irrotational flows and bodies in cases of *unsteady* fluid motion (those forces are, of course, zero in steady motion as shown in Section 7.4).

Although the results of this section, including eqn (305), have important applications in a wide range of different types of motion, we here apply them solely to the case of irrotational flows set up by the pure translational movement of a solid body with velocity $\mathbf{U}$. The boundary condition (eqn (74)) satisfied by the irrotational flow on the body surface S is then

$$\partial\phi/\partial n = \mathbf{u}\cdot\mathbf{n} = \mathbf{U}\cdot\mathbf{n}. \tag{308}$$

This condition allows us to obtain a very simple expression for the total kinetic energy of the flow field, a concept which, unlike the 'total momentum', has an unambiguous meaning because the kinetic energy per unit volume $\frac{1}{2}\rho(\mathbf{u}\cdot\mathbf{u})$ falls off as fast as r^{-6} as $r\to\infty$, so that its volume integral converges. In fact, we may extend the argument of eqn (295) to the case when $\mathbf{u}\cdot\mathbf{n}$ takes the value in eqn (308) (instead of being zero) on S to deduce that the kinetic energy of the fluid in the region L between S and a large sphere Σ_R is

$$\frac{1}{2}\rho\int_{L}(\mathbf{u}\cdot\mathbf{u})\,dV = \frac{1}{2}\rho\int_{\Sigma_R}\phi\mathbf{u}\cdot\mathbf{n}\,dS + \frac{1}{2}\rho\int_{S}\phi\mathbf{U}\cdot\mathbf{n}\,dS, \tag{309}$$

where the first integral on the right-hand side of eqn (309) tends to zero as R becomes large exactly as shown for eqn (295). Therefore, the total kinetic energy E of the irrotational flow is given by the last integral in eqn (309), which by eqn (304) may be written

$$E = \frac{1}{2}\mathbf{U}\cdot\mathbf{I}. \tag{310}$$

We may note that eqn (310) represents the relationship between impulse and energy that might have been expected from various general considerations. For example, if the impulse (in eqn (170)) were applied through the action of a *constant* very large force $\mathbf{F}$ over the very short time $0 \leqslant t \leqslant \tau$, during which the body velocity increased steadily like $(t/\tau)\mathbf{U}$, the *rate* of working by the body on the fluid would be $(t/\tau)\mathbf{U}\cdot\mathbf{F}$ and the resulting kinetic energy transmitted to the fluid would be

$$\int_{0}^{\tau}(t/\tau)\mathbf{U}\cdot\mathbf{F}\,dt = \frac{1}{2}\mathbf{U}\cdot\mathbf{F}\tau = \frac{1}{2}\mathbf{U}\cdot\mathbf{I}. \tag{311}$$

Again, the principles of general analytical dynamics are consistent with the eqn (310) for kinetic energy if we regard the components of $\mathbf{U}$ as 'generalized velocities' determining the irrotational motion as a dynamical system.

With the basic principle (eqn (305)), eqn (310) tells us that

$$\frac{1}{2}\rho\mathbf{U}\cdot\mathbf{G} = E + \frac{1}{2}\rho V U^{2}; \tag{312}$$

where the right-hand side is necessarily positive, being the sum of the kinetic energies of (i) the irrotationally moving fluid and (ii) the hypothetical 'displaced fluid' of mass ρV and velocity $\mathbf{U}$. This demonstrates that the far-field dipole strength $\mathbf{G}$ is necessarily non-zero and, indeed, that it necessarily has a *positive* component in the direction of the body velocity $\mathbf{U}$.

Similarly, the requirement that eqn (310) must be positive shows that $\mathbf{I}$ must have a positive component in the direction of $\mathbf{U}$. There are, in fact, many cases when considerations of symmetry clearly show that **I** is exactly aligned with **U**, and we confine ourselves to such cases in the rest of this section while at the same time mentioning that other cases involving a non-zero angle between the vectors **I** and **U** will be studied in Section 8.4.

For example, any *axisymmetric* body, like the Rankine ovoid of Section 7.5, must (by reasons of symmetry) exert upon the fluid an impulse **I** directed along the axis when it instantaneously sets up the irrotational motion associated with its own movement *along the axis* with velocity **U**. In such a case, then, we can write simply

$$\mathbf{I} = m_a \mathbf{U} \tag{313}$$

in terms of a scalar coefficient m_a. Evidently, this coefficient is independent of the magnitude U of the body velocity **U**, since the solution ϕ of Laplace's equation which satisfies eqn (308) on S must increase linearly with U so that the impulse (eqn (304)) has the same property.

The coefficient m_a in eqn (313) is usually called the 'added mass' (or, sometimes, 'virtual mass'). One reason for the name is that, if the body itself is of mass m, it certainly requires an externally applied impulse $m\mathbf{U}$ to set it instantaneously into motion with velocity **U** even if there be no fluid surrounding the body to impede its acceleration. But, with the *fluid* needing an additional impulse $m_a\mathbf{U}$ to set it into irrotational motion, the total externally applied impulse must be

$$(m + m_a)\mathbf{U}. \tag{314}$$

Thus, the presence of the fluid effectively imparts an additional inertia or added mass m_a to a body of mass m immersed within it.

Similarly, the work which must be done to set the immersed body into motion with velocity **U** is the sum of the body's kinetic energy $\frac{1}{2}mU^2$ and the fluid kinetic energy as given by eqns (310) and (313). Thus, it is

$$\tfrac{1}{2}(m + m_a)U^2, \tag{315}$$

exactly as if the body had effectively possessed a total mass $m + m_a$.

As yet another illustration, the external *force* required to give the immersed body of mass m an acceleration $d\mathbf{U}/dt$ consists of its normal value $md\mathbf{U}/dt$ plus the additional force $\mathbf{F} = m_a d\mathbf{U}/dt$, specified by eqns (306) and (313), which the body must apply to the fluid. This gives the total force as

$$(m + m_a)d\mathbf{U}/dt, \tag{316}$$

just as if the mass m had been increased by the added mass m_a.

In terms of the magnitude G of the far-field dipole strength **G**, the principle in eqn (305) allows us to write the added mass defined in eqn (313) as

$$m_a = \rho(U^{-1}G - V), \tag{317}$$

where V is the volume of the body. For example, a sphere of radius a has

$$V = \tfrac{4}{3}\pi a^3, \quad \text{and} \quad G = 2\pi U a^3 \tag{318}$$

by eqn (251), so that the added mass (expression (316)) must be

$$m_a = \rho(\tfrac{2}{3}\pi a^3) = \tfrac{1}{2}\rho V, \tag{319}$$

namely half the mass of the displaced fluid.

Actually, for a shape as simple as a sphere, the correctness of this eqn (319) is readily checked by calculating the resultant of the *transient pressures* $-\rho \partial \phi / dt$, where ϕ is given by eqn (257), over the surface of the sphere as $m_a dU / dt$ with m_a given by eqn (319). (Only the transient pressures need be considered because the drag in steady irrotational flow is known to be zero.) With more complicated shapes, however, a calculation based on eqn (317) is often much more straightforward.

For a solid sphere immersed in *air*, the added mass (eqn (319)) may represent a negligible addition to the normal mass of the sphere. If the sphere is immersed in water, however, the addition becomes significant, although our formulas such as expression (314) apply *only* in the very earliest stages of the sphere's motion before the drag associated with formation of a vortex wake becomes significant.

For a spherical *bubble* in water, furthermore, the added mass $\tfrac{1}{2}\rho V$ given by eqn (319) may dominate completely over the much smaller mass of the bubble itself. Initially, then, the upward buoyancy force $\rho V g$ acting on the bubble, with its total effective mass of $\tfrac{1}{2}\rho V$, can cause it to move upwards with an acceleration $2g$.

In the *other* extreme case of the Rankine ovoid, where the distance $2h$ between the source and the sink (of equal strengths J) is large compared with the radius $a = (J/\pi U)^{\frac{1}{2}}$ of the Rankine fairing, we have seen (Fig. 48) that the shape of each half of the ovoid is essentially that of the Rankine fairing itself. We then calculate that the ovoid's volume V is

$$V = 2(h\pi a^2 - \tfrac{1}{3}\pi a^3), \tag{320}$$

where the first term in brackets is the volume of a circular cylinder of radius a stretching between $z = -h$ and $z = 0$, and the second term represents a correction, obtained from the shape (eqn (233)) of the Rankine fairing as

$$\int_0^a 2\pi s z \, ds = \tfrac{1}{3}\pi a^3. \tag{321}$$

In the far field, on the other hand, the source and sink of strength J at a distance $2h$ apart must behave like a dipole of strength

$$G = 2hJ = 2h\pi U a^2. \tag{322}$$

Therefore, the added mass (expression (316)) is given, by eqns (320) and (322), as

$$m_a = \rho(\tfrac{2}{3}\pi a^3).\tag{323}$$

For the slender, cigar-shaped (and relatively streamlined) body of Fig. 48, then, the ratio of the added mass m_a to the mass ρV of displaced fluid has a value $a/(3h-a)$ which is *far* smaller than the value $\tfrac{1}{2}$ already noted for spheres in eqn (319). Such slender shapes, in fact, are relatively much more easy to accelerate initially in fluid, as well as more easy to push through fluid after they have got up speed.

8.4 Moments applied by steady irrotational flows

The symmetry of a sphere is such that its added mass must obviously take the same value (calculated in eqn (319) as $\tfrac{1}{2}\rho V$) when it moves in any direction whatsoever. We have, however, no reason to expect a similar directional independence of added mass for bodies lacking spherical symmetry. In this section, we first establish that fluid inertia may show a pronounced directional dependence; a property which we then, in particular, quantitatively relate to the fact that although steady irrotational flow around a solid body cannot exert any force on that body, it may apply a moment or torque to that body.

A good example of a strong directional dependence of added mass is provided by the slender, cigar-shaped Rankine ovoid of Fig. 48. When it moves along its axis, the associated added mass m_a takes a small value (eqn (323)), equal to half the mass of a fluid sphere whose radius a is the maximum radius of the ovoid. We can understand this small value of m_a in terms of the kinetic energy $\tfrac{1}{2}m_a U^2$ of the fluid. The axial motion of the ovoid with velocity U generates fluid velocities equal to U at its two ends, but the region of fluid within which the velocities generated are comparable to U is restricted to within a distance of the order of a or less from either end (Fig. 54), which limits the fluid's kinetic energy to a value of order $\rho a^3 U^2$ (the actual value being $\tfrac{1}{3}\pi\rho a^3 U^2$).

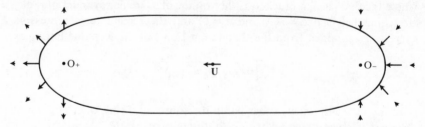

Fig. 54. Typical fluid motions responsible for the added mass of a Rankine ovoid in axial motion through otherwise undisturbed fluid.

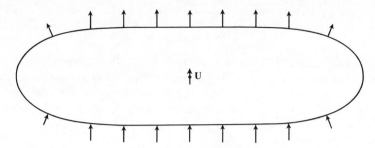

Fig. 55. Typical fluid motions responsible for the much greater added mass of a Rankine ovoid in transverse motion through fluid.

By contrast, if the ovoid were to move at right angles to its axis (Fig. 55), fluid velocities comparable to U would be produced throughout a region of fluid whose length is approximately that of the body while it lateral dimensions are of the order of the body radius. This suggests the possibility of a fluid kinetic energy comparable to $\frac{1}{2}\rho V U^2$ (where V is the volume of the body), which would correspond to an added mass comparable to ρV, enormously larger than the added mass (from eqn (323)) for axial motion of the body.

We now give a more quantitative (albeit only roughly approximate) analysis tending to support the closeness of the above estimates. Evidently, the ovoid's *fore-and-aft symmetry* (invariance to reflection about the plane $z = 0$) implies that the impulse **I** exerted upon the fluid when the ovoid acquires a velocity **U** perpendicular to its axis must be in the same direction as **U**, leading as in eqn (313) to an equation $\mathbf{I} = m_a \mathbf{U}$ which defines the added mass in this case. We proceed to determine m_a from eqn (317) in terms of the magnitude G of the far-field dipole strength.

This determination can be carried out to a rough approximation for a *slender* Rankine ovoid by using the idea that the form of eqn (294) of the dipole potential fits it peculiarly well for satisfying the boundary condition (in eqn (308)) on the surface of bodies moving at right angles to an axis of rotational symmetry. This categorization includes, of course, the sphere (end of Section 7.5) for which every axis through the centre is one of symmetry, but the same conclusion applies to the Rankine ovoid of Fig. 55 and most obviously to its central cylindrical section.

With the direction of motion of the ovoid taken as that of the x axis, so that $\mathbf{U} = (U, 0, 0)$, and with equations

$$x = s\cos\psi, \quad y = s\sin\psi \qquad (324)$$

accompanying eqns (226) to complete the relationship of the Cartesian coordinates x, y, z to the cylindrical coordinates z, s, ψ, the boundary condition (eqn (308)) takes the form

$$\partial\phi/\partial s = U\cos\psi \tag{325}$$

on the ovoid's central cylindrical portion $s = a$. It is easy to satisfy this condition, not as with a sphere by placing a single dipole at its centre, but by means of a continuous distribution of dipoles along the axis $s = 0$ of the cylinder. With the strength of the dipoles between $z = Z$ and $z = Z + dZ$ taken as

$$\mathbf{G} = (g\,dZ, 0, 0), \tag{326}$$

the velocity potential (eqn (294)) for these dipoles is

$$-\frac{g\,dZ}{4\pi}\frac{x}{[x^2 + y^2 + (z - Z)^2]^{\frac{3}{2}}} = -\frac{g\,dZ}{4\pi}\frac{s\cos\psi}{[s^2 + (z - Z)^2]^{\frac{3}{2}}}, \tag{327}$$

and at a point on the cylinder $s = a$ the combined effect of such dipoles (eqn (327)) distributed over positions Z extending to at least two or three radii on either side of z is close to the values obtained by integrating eqn (327) from $Z = -\infty$ to $Z = +\infty$. This is

$$\phi = -\frac{g\cos\psi}{2\pi s}, \tag{328}$$

which satisfies eqn (325) on $s = a$ if the dipole strength per unit length along the axis is

$$g = 2\pi U a^2. \tag{329}$$

Interestingly, $U^{-1}g$ is *twice* the value (πa^2) of the body volume per unit length along the axis. Although this conclusion has been derived only for the central cylindrical portion of a Rankine ovoid, it is nevertheless found to remain true for such a slender, axisymmetric body as a whole that a boundary condition (eqn (325)) appropriate to its motion in the x direction, at right angles to its axis, can be satisfied by means of a distribution (eqn (326)) of dipoles along the axis with $g = g(Z)$ varying with Z in such a way that the total dipole strength $G = \int g(Z)dZ$ has $U^{-1}G$ rather close to $2V$ (twice the volume of the body); with the closeness of approximation being improved as the body becomes more and more slender (as when, for the Rankine ovoid, a/h becomes smaller).

In the far field, of course, this entire distribution of dipoles is 'seen' as a single dipole of strength G, with

$$U^{-1}G \doteqdot 2V. \tag{330}$$

By eqn (317) this gives an approximate determination of the added mass as

$$m_a \doteqdot \rho V, \tag{331}$$

an equation stating that the added mass, in this case, is approximately *equal* to the mass of the displaced fluid.

Although the above argument is very crude, and the conclusion (eqn (331)) is in fact only quite a rough approximation, it nevertheless leaves us in no doubt that the added mass for motion of a slender Rankine ovoid at right angles to its axis is vastly greater than the quite small value in eqn (323) which has been obtained for the added mass associated with the ovoid's axial motion. The rest of this section probes some consequences of this conclusion.

Using the notations m_x and m_z for the values of added mass associated with motion in the x and z directions respectively, we consider a motion of the body with velocity

$$\mathbf{U} = (U_x, 0, U_z).\tag{332}$$

The impulse that must act on the fluid if the body is to acquire the velocity $(U_x, 0, 0)$ is $(m_x U_x, 0, 0)$, whereas an impulse $(0, 0, m_z U_z)$ is needed for it to acquire the velocity $(0, 0, U_z)$. The combined velocity (eqn (332)) requires, then, both these impulses to be applied (whether in immediate succession or simultaneously) to give a total

$$\mathbf{I} = (m_x U_x, 0, m_z U_z).\tag{333}$$

This impulse is not in general parallel to the velocity vector in eqn (332). (In parentheses, we may note the form

$$E = \tfrac{1}{2}\mathbf{U}\cdot\mathbf{I} = \tfrac{1}{2}m_x U_x^2 + \tfrac{1}{2}m_z U_z^2\tag{334}$$

for the kinetic energy (eqn (310)) in this case.)

The fact that $\mathbf{I}$ need not be parallel to $\mathbf{U}$ has a most interesting consequence in the case of a finite body moving steadily (that is, with *constant* velocity $\mathbf{U}$) through externally unbounded fluid. In this case, on the assumption of irrotational flow, eqn (333) makes the impulse $\mathbf{I}$ constant so that the total force $\mathbf{F}$ applied to the fluid is given by eqn (306) as

$$\mathbf{F} = d\mathbf{I}/dt = 0.\tag{335}$$

We can, however, look again at the demonstration of eqn (306) to show in the present case that, although it correctly predicts a system of forces with zero resultant in eqn (335), those forces are nevertheless equivalent to a couple with, in general, non-zero *moment*.

This conclusion is derived from the fact that the impulse $\mathbf{I}$ required to set up the motion of a body with velocity $\mathbf{U}$ must have a certain *line of action*, a point $\mathbf{x}_1$ on the line of action having the property that the moment of the impulse about $\mathbf{x}_1$ is zero.[†] Now, after a time dt, both the body's motion with velocity $\mathbf{U}$,

---

[†] For example, with the Rankine ovoid's fore-and-aft symmetry, its geometric centre is necessarily such a point.

and the associated irrotational motion of the fluid, are exactly the same as before although the body has received a vector displacement $\mathbf{U}dt$ so that the whole pattern of fluid motion has also been displaced by this amount. In particular, the line of action of $\mathbf{I}$ now passes through

$$\mathbf{x}_1 + \mathbf{U}dt. \tag{336}$$

Accordingly, the system of forces with which the body acts on the fluid during time dt must produce an impulsive effect equivalent to generating the impulse $\mathbf{I}$ through the point described by expression (336) while generating an equal and opposite impulse $(-\mathbf{I})$ through $\mathbf{x}_1$ to cancel the former flow's impulse. The effect is equivalent, then, to an *impulsive couple*, that is two equal and opposite impulses with zero resultant, whose moment about $\mathbf{x}_1$ is readily calculated as the vector product

$$(\mathbf{U}dt) \times \mathbf{I}; \tag{337}$$

and, because the resultant is zero, this must also be the moment of the system of impulses about any point. The system of *forces*, therefore, must have a moment obtained through division by dt as

$$\mathbf{M} = \mathbf{U} \times \mathbf{I}. \tag{338}$$

Thus, whenever the impulse $\mathbf{I}$ is not parallel to the velocity $\mathbf{U}$, a non-zero moment (eqn (338)) must be applied to the fluid by the pressures acting at the body surface. To every action, of course, there is an equal and opposite reaction, so that the fluid pressures necessarily apply a moment $(-\mathbf{M})$ to the body. This moment has to be withstood by an external couple $\mathbf{M}$ if the body is to continue in steady motion. This means that, although no external force or moment need be applied to maintain a body in steady translational motion *in vacuo*, an external couple of moment $\mathbf{M}$ does need to be applied to maintain such a motion within a fluid moving irrotationally. For example, when eqns (332) and (333) represent $\mathbf{U}$ and $\mathbf{I}$, eqn (338) gives

$$\mathbf{M} = (0, (m_x - m_z)U_x U_z, 0). \tag{339}$$

For a body such as the slender Rankine ovoid the above results are of practical importance in the case when the body moves steadily at a *small* angle α to its axis of symmetry, making only a small perturbation of the axisymmetric flow regime for which (Fig. 48) the irrotational-flow pressure distribution is quite closely realized over almost all the body surface. Then, with $U_x = U \sin \alpha \doteqdot U\alpha$ and $U_z = U \cos \alpha \doteqdot U$, and with $m_x - m_z \doteqdot \rho V$, the moment $(-\mathbf{M})$ applied by the fluid pressures takes a value of approximate magnitude $\rho V U^2 \alpha$ which tends to *increase* α.

This means that the axial motion of such a slender axisymmetric body is necessarily unstable: any small angle α between its direction of motion and its axis tends to increase exponentially as a result of the fluid applying a couple directly proportional to α. This is why any such body needs in practice to be fitted with *fins* at its rear end (Fig. 56) which, for reasons that will be appreciated later, exert an opposing couple tending to stabilize the motion of the body.

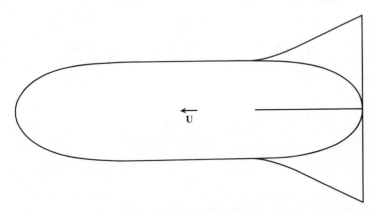

Fig. 56. Because a slender axisymmetric body is subjected to a destabilizing moment tending to increase any small positive angle α between its direction of motion and the axis of symmetry, stabilizing *fins* are often fitted at the rear ends of such bodies (see also Section 11.4).

The title of this section was chosen to help remind readers that all results for a finite body moving steadily through unbounded fluid which is at rest far from the body can be alternatively viewed in a frame of reference in which the body is maintained at rest within a uniform stream of fluid. In that frame of reference our conclusions are that, although pressures in the steady irrotational flow exert no force on the body (eqn (335), confirming and extending the earlier conclusion (eqn (240)) of Section 7.4), they do nevertheless apply to it a couple of moment $(-\mathbf{M})$ so that an external couple of moment $\mathbf{M}$ must be applied to maintain the body at rest.

8.5 Drag with a wake

Next we consider the more realistic case of steady flow around a body which is irrotational only outside the boundary layer and associated wake and on which, in consequence, a positive drag force D acts on the body. The far-field properties are modified in this case in a way directly related to the value of D.

Throughout the irrotational part of the flow, eqn (179) assures us that $\frac{1}{2}q^2 + \rho^{-1}p_e$ must take a single constant value, which must be

$$\frac{1}{2}q^2 + \rho^{-1}p_e = \frac{1}{2}U^2 \qquad (340)$$

if U is the value of the fluid speed q in the uniform stream where $p_e = 0$. The wake, however, consists of streamlines on which dissipative actions (associated with viscosity and, usually, turbulence[†]) has been taking place so that, as noted in connection with eqn (84), the 'total head' has failed to remain constant along those streamlines, having (in the main) been *reduced* by energy dissipation (just as in cases noted in Section 1.3). In the wake, then,

$$\tfrac{1}{2}q^2 + \rho^{-1}p_e < \tfrac{1}{2}U^2. \tag{341}$$

However, the result in expression (87) implies that the excess pressure p_e cannot vary significantly across the approximately straight streamlines of the wake. Far from the body, then, the wake is a region where $p_e = 0$ and where

$$q < U. \tag{342}$$

This represents a definite reduction of fluid speed within that narrow region (Fig. 57).

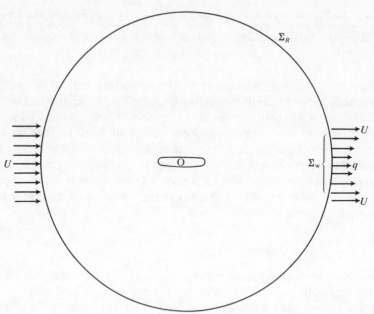

Fig. 57. Far field analysis for a body with a wake. The large sphere Σ_R of centre O and radius R cuts the wake in a region Σ_w where reduced total head (due to viscous, and turbulent, dissipation and diffusion) causes the fluid speed q to take values less than the speed U of the oncoming flow.

[†] If the wake is turbulent, we must take q to signify the time-averaged fluid speed at a point.

A large sphere Σ_R whose radius is R and whose centre O is within the body cuts the wake in a small slice Σ_w (Fig. 57) where the property in the inequality (342) implies a *defect of volume flow*

$$J = \int_{\Sigma_w} (U - q)\,\mathrm{d}S, \tag{343}$$

across the slice. The incompressibility of the fluid requires this to be made up over the rest of the sphere Σ_R; where, however, the fluid motion relative to the uniform stream is irrotational so that its far field must take the general form of eqn (290) which was derived in Section 8.2. We deduce that the value of the source strength J for that far field must be as given by eqn (343) in order that the excess outward volume flow in the irrotational part of the far field,

$$J = \int_{\Sigma_R} (\partial\phi/\partial n)\,\mathrm{d}S, \tag{344}$$

shall balance the defect of volume flow in the wake.

Since the far-field source strength (eqn (344)) is a constant independent of R, we deduce that the wake's defect of volume flow (eqn (343)) is also independent of R. This means that, as diffusion of vorticity gradually causes the wake to become thicker, it must do so in such a way that the integral (eqn (343)) remains constant.

As before, any study of the *force* acting between the body and the fluid is best carried out in a frame of reference in which the body is moving with constant speed and the fluid far from it is at rest. Between time t and $t + \mathrm{d}t$ no net impulse needs to be applied to change the irrotational part of the flow in this frame of reference (since, as in eqn (335), it has a constant impulse I). However, the wake in this frame of reference has a momentum ρJ per unit length in the direction of motion of the body, where J is given by eqn (343). During time $\mathrm{d}t$, the wake lengthens by a further amount $U\mathrm{d}t$ so that an additional momentum $\rho J(U\mathrm{d}t)$ has to be supplied to it, requiring the action of a force

$$D = \rho U J. \tag{345}$$

This force represents the thrust which must be applied to the body to push it through the fluid, and it is of course equal in magnitude and opposite in direction to the drag D with which the fluid pressures and viscous stresses act on the body.

A body that leaves a wake, then, has an irrotational flow far field outside the wake that includes a source term in which the source strength J takes a value $D/\rho U$ directly proportional to the drag on the body. Even for a body with rather small drag D that generates an irrotational flow with quite a large dipole strength G, the source term in its far field may ultimately become dominant at very large distances r from the body.

These considerations need to be taken into account in estimating any effect on the fluid motion exercised by *distant boundaries*. Real fluids, of course, are never externally unbounded in a completely literal sense; nevertheless, if the fluid's boundaries are distant enough, the model of an externally unbounded fluid may be very useful. Furthermore, this model is useful not only for giving a first approximation to the fluid motion but also for going to a second approximation that takes into account modifications due to those boundaries.

Suppose, for example (Fig. 58), that the nearest external boundary of the fluid to the body (and, therefore, the external boundary most likely to have a significant effect) is a flat ground under it. Then the first approximation to the flow, given when we model it as externally unbounded, has the same far field (eqn (290)) as a source and/or a dipole situated at the origin O. However, at a solid plane boundary (representing the ground), this far field does not satisfy the necessary boundary condition $\mathbf{u} \cdot \mathbf{n} = 0$.

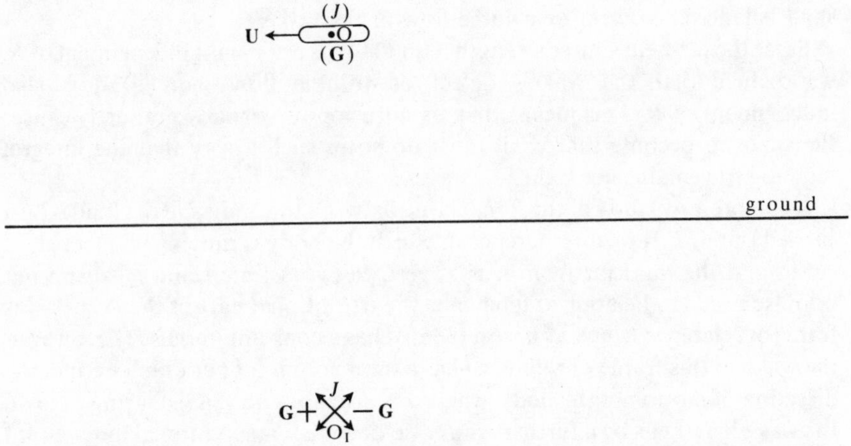

Fig. 58.   Flow due to motion of a body of centroid O at velocity U in fluid bounded below by flat ground. The far field in unbounded fluid would be that of a dipole G situated at O together with that of a source J related to the body's wake. The addition of an equal dipole and source at the image point O_I allows the boundary condition at the ground to be satisfied.

Fortunately, we know already, from Fig. 37, how to remedy this situation. We add the *mirror-image* of the far-field flow (see Fig. 58, where an image source and an image dipole are placed at the point O_I which is the mirror-image of O in the plane) so that the sum of that flow and its mirror-image has zero normal velocity on the plane. If now we calculate that the velocity field of the mirror-image source/dipole system takes the value $\mathbf{U}_I$ at O, then this is also a good estimate of its value throughout the *neighbourhood* of the body (because the latter's size is small compared with its distance from O_I).

Therefore, to a second approximation we can calculate the fluid motion in the neighbourhood of the body as that of an externally unbounded fluid with the velocity **U** of the body relative to the undisturbed motion of the fluid replaced by $\mathbf{U} - \mathbf{U}_I$.

Similar methods may commonly be used with other shapes of distant boundary. Often, they involve multiple images; and, for a body being tested in a wind-tunnel of rectangular cross-section, they involve a doubly infinite array of images (Fig. 59). This, we may note, turns out to be a case when the presence of a source of strength J in the image-system makes a special modification of the irrotational flow outside the wake so as to generate a difference J/A between its downstream and upstream fluid speeds (where A is the tunnel's cross-sectional area). This modification occurs additionally to any changes in the effective vector wind velocity in the body's neighbourhood which is associated with the array of dipole images.

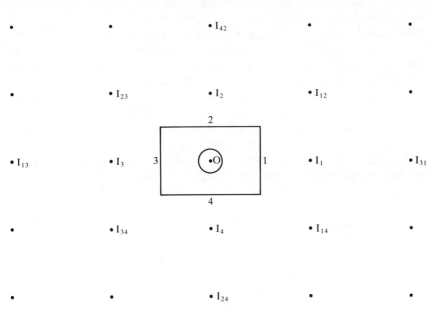

Fig. 59. Cross-section of a wind-tunnel in which an axisymmetric body is being tested. Here, a *doubly infinite* array of images is needed. For example, the points $\mathbf{I}_m$ (for $m = 1$ to 4) are the images of O in the wind-tunnel's sides 1, 2, 3 and 4, while $\mathbf{I}_{mn}$ is the image of $\mathbf{I}_m$ in side n for each $n \neq m$, and so on.

9
Two-dimensional irrotational flows

As remarked in Section 7.1, a specially convenient analytical method for obtaining solutions to Laplace's equation (eqn (208)) exists in cases when the third term on the left-hand side is of negligible importance so that, to good approximation, ϕ satisfies the two-dimensional Laplace equation

$$\partial^2 \phi / \partial x^2 + \partial^2 \phi / \partial y^2 = 0. \tag{346}$$

That method, based on the theory of the *complex* potential and the technique known as conformal mapping, is introduced in the present chapter.

Although the two-dimensional eqn (346) is mathematically simpler than the full Laplace equation (eqn (208)), and the analytical method referred to facilitates the derivation of interesting solutions in abundance, no temptation to embark upon two-dimensional theory in advance of the general three-dimensional theory of chapter 6, 7 and 8 could have been justified by these advantages. Indeed, the advent of fast and convenient computer algorithms for solving both eqns (208) and (346) has in any case reduced the special advantage of eqn (346) which results from the better availability of analytical solutions, even though it has by no means reduced the need to understand the fundamentals of the theory of conformal mapping and the complex potential. A much more important consideration, however, is that the physical interpretation of solutions of the two-dimensional eqn (346) is relatively more difficult, and needs the background of the general three-dimensional theory to make it intelligible. In many of the most useful two-dimensional solutions, for example, the kinetic energy of the disturbances may be infinite; it may in addition be impossible to set up the flow from rest by impulses acting at its boundaries; yet, if we regard the motion as locally a two-dimensional irrotational flow that nevertheless is embedded in a more complex motion which is three-dimensional or rotational (or both), then its relevance becomes intelligible.

9.1 Two-dimensional regions embedded within three-dimensional flows

One of the greatest benefits to theoretical fluid mechanics from two-dimensional analysis lies in an elucidation of the properties of the *aerofoil* (spelt as 'airfoil' in the US literature). This word is generally used to mean the *cross-section, at right angles to its length,* of any long surface that is well adapted to generate forces in a direction nearly perpendicular to its direction of motion relative to undisturbed fluid; that is, of a surface such as an aircraft wing (a) and (e), a turbine blade (b), a dolphin's tail fluke (c), or a shark's tail fin (d). Figure 60 illustrates typical cross-sections for each of these cases, characterized by a well rounded *leading edge* at the front end (where the aerofoil moves into the fluid) and a sharply pointed *trailing edge* at the rear end.

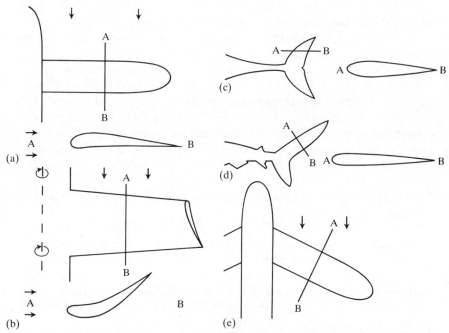

Fig. 60. Schematic illustration of various aerofoil sections in engineering and nature. In each case the cross-section AB of the wing or winglike surface has the aerofoil shape shown. The arrows indicate the motion of fluid relative to the surface.

(a)   An aircraft's unswept wing, with a moderately *cambered* aerofoil section (see Section 10.3).

(b)   The rotating blade of a gas turbine, with its aerofoil section much more cambered (see Section 11.4).

(c)   The tail fluke of a dolphin with a symmetrical aerofoil section.

(d)   The sweptback caudal fin of a relatively slow-moving species of shark (the blue shark) with a thinner symmetrical aerofoil section.

(e)   Modern aircraft have sweptback wings for a different reason (Section 11.4), a typical aerofoil section being broadly similar to that shown under (a) above.

For any long, winglike surface of this type, preliminary insight into the local action of a particular aerofoil section on the fluid can be obtained by treating the resulting motion as a two-dimensional flow in the plane of the aerofoil section. This procedure depends on the idea that rates of change along the *length* of the winglike surface must be much smaller than rates of change in the plane (at right angles to that length) of the aerofoil section itself. Thus, choosing x and y coordinates in that plane and the z direction along the length of the winglike surface, we have

$$\partial^2 \phi / \partial z^2 \ll \partial^2 \phi / \partial x^2 \text{ or } \partial^2 \phi / \partial y^2, \tag{347}$$

which justifies the approximate replacement of eqn (208) by eqn (346).

The solution of eqn (346) which is chosen is determined by the relative velocity between the aerofoil section and the undisturbed fluid, *resolved in the plane of the section*. Thus, in a simple case such as the straight wing of Fig. 60 (a) in forward flight, it depends on the velocity of the wing relative to the undisturbed air. This velocity, in magnitude and direction, determines the choice of two-dimensional irrotational flow around the aerofoil section.

For the sweptback wing of Fig. 60 (e), on the other hand, it is only a part of the wing's velocity (the part resolved in the plane of the aerofoil section) which determines that two-dimensional motion of the air which results from an interaction between the aerofoil's shape and its motion in its own plane. By contrast, the z component of the wing's velocity (that is, the component along the length of the wing) may be regarded as producing no significant disturbance to the air. Similar remarks apply to the sweptback shark's fin of Fig. 60(d); see Chapter 11 for comments on the quite different advantages of sweepback in these two cases.

Within the plane of the aerofoil section, the motion is determined as that of a fluid which is bounded internally by the aerofoil and which, just as in Chapter 8, is externally unbounded. Although it is the aerofoil's own movement relative to the undisturbed fluid which excites the motion, that motion is usually analysed in a frame of reference in which the aerofoil is held at rest in a uniform stream of fluid, the velocity of this stream being, of course, equal and opposite to that of the aerofoil relative to the fluid.

Such a mathematical model of the two-dimensional flow of a uniform stream about an aerofoil section can be directly compared with experiment in a wind-tunnel. For this purpose a long test specimen with uniform cross-section (that of the aerofoil shape in question) is placed at right angles to the wind, with arrangements being made, usually, for the specimen to be turned about its long axis so that it can be tested at various *angles of incidence* (sometimes called 'angles of attack') to the oncoming wind.

Some problems special to the two-dimensional theory of irrotational flow around an aerofoil section arise from the fact that the fluid region outside the aerofoil is *doubly-connected*. This implies (Section 6.4) that there is no unique flow satisfying the boundary conditions. Instead, there is a wide range of such flows, and the flow is determined uniquely if and only if a prescribed value is given to the circulation K around a certain curve embracing the aerofoil. To be sure, just one particular motion results if it is excited impulsively from rest

(Section 6.1): a motion which necessarily has a velocity potential, given by eqn (173), so that the circulation K is zero in this one case. None of the other motions, with $K \neq 0$, can possibly have a velocity potential in the usual sense of a continuous single-valued function with grad $\phi = \mathbf{u}$ (see eqns (202) and (203), for example). Nevertheless, careful analysis of the aerofoil problem in Chapter 10 shows that the particular solution which is justified by that analysis, and which is relevant to the aerofoil's capability of generating forces perpendicular to the stream, is one with a non-zero value of the circulation K. This motion is important in spite of the apparently anomalous feature that it possesses infinite energy. Furthermore, this two-dimensional solution is valuable not only as a first approximation to the local motion (Chapter 10) but also as a component of a wider theory for the winglike surface as a whole (Chapter 11) which allows us to obtain a second approximation to the local fluid motion as a two-dimensional solution associated with a slightly modified value of the relative velocity between the aerofoil and the neighbouring fluid.

Postponing until Chapter 10 any general analysis of flows with non-zero circulation, and until Chapter 11 any general account of the embedding of two-dimensional aerofoil flows within a three-dimensional wing theory, we concentrate in Chapter 9 on the fundamentals of two-dimensional irrotational flow theory; concluding this introductory section with two very simple examples of problems special to that theory which demonstrate a need for it to be related to some wider three-dimensional considerations. Both examples are additionally important as giving simple illustrations of *singular solutions* of the two-dimensional Laplace equation, a type of solution which we know from Chapters 7 and 8 to be potentially very valuable.

One important two-dimensional flow that is irrotational everywhere except at an isolated singularity is the *line vortex* (see Figs. 30 and 39). The velocity field is two-dimensional since its direction lies in a plane perpendicular to the line vortex and its distribution in each such plane is the same, with magnitude $q = K/(2\pi r)$ directed tangentially to a circle of radius r centred on the vortex.

Nevertheless, this flow exhibits two anomalies. First, it possesses no velocity potential in the usual sense, as Fig. 39 shows (thus, it cannot be set up from rest by impulsive action at its boundaries). Admittedly, the usual integral defining a velocity potential can be written down as in eqn (202) but it is not a continuous one-valued function; it can be made one-valued by the familiar device of restricting θ to the interval $-\pi < \theta \leqslant \pi$ but then it is discontinuous at the radius $\theta = \pi$ so that the 'candidate velocity potential'

$$\phi = K\theta/(2\pi) \tag{348}$$

satisfies grad $\phi = \mathbf{u}$ everywhere *except* on that radius.

A second difficulty is associated with the energy of this line-vortex flow. For any two-dimensional flow, the *kinetic energy per unit span* (that is, per unit length in the z direction) can be written as an integral

$$\int_A \tfrac{1}{2} \rho q^2 \, dx \, dy \tag{349}$$

over the area A of the (x, y) plane that is occupied by fluid. If, as in Fig. 39, this area extends from $r = r_1$ to $r = r_2$ then the integral (expression (349)) can be written

$$\int_{r_1}^{r_2} (\tfrac{1}{2} \rho \, q^2) \, 2\pi r \, dr, \tag{350}$$

which with $q = K/2\pi r$ takes the form

$$(\rho K^2 / 4\pi) \int_{r_1}^{r_2} r^{-1} \, dr = (\rho K^2 / 4\pi) \log (r_2/r_1). \tag{351}$$

This becomes infinite if *either* $r_1 = 0$ *or* $r_2 = \infty$. The first anomaly is easily explained since a line vortex is always an idealization of a bundle of vortexlines of definite thickness, so that r_1 can never properly be reduced to zero. The second anomaly indicates, however, that calculations for externally unbounded fluid may be extended to the two-dimensional case only with caution.

One further example of a singular solution demonstrates that this particular difficulty is by no means confined to flows with circulation. The two-dimensional analogue of a point source is the so-called *line source*, formed by a continuous and uniform distribution of point sources of *strength j per unit span* (that is, per unit length of a straight line parallel to the z direction). Thus, the rate of volume outflow from the line source is j per unit span, which requires the radially outward velocity at a distance r from the source to be of magnitude

$$q = j/(2\pi r). \tag{352}$$

This line-source flow does possess a well defined velocity potential

$$\phi = (j/2\pi) \log r \tag{353}$$

whose gradient is directed radially outwards (in the direction r increasing) and has magnitude as in eqn (352). However, the kinetic energy (the value in expression (350) per unit span) takes exactly the same form as in eqn (351) with j replacing K so that, once again, it becomes infinite in the case $r_2 = \infty$ of externally unbounded fluid. In particular, any far field of source type (analogous to those of Chapter 8 in the three-dimensional case) would have infinite energy.

This anomalous feature of the line source can be eliminated only through embedding the two-dimensional flow (eqn (353)) within a three-dimensional motion associated with sources distributed along a straight line of finite length. In the case of the line vortex too, such an embedding is necessary, but it must take into account the solenoidality of the vorticity field which requires a line vortex of finite length to be part of a continuous closed loop of concentrated vortex lines (see Chapter 11). In the rest of Chapter 9, however, we put aside questions of how the two-dimensional flows we consider may fit into wider three-dimensional patterns, and concentrate on the mathematical techniques for deriving useful solutions to the two-dimensional Laplace equation (eqn (346)).

9.2 The complex potential

Our methods are based on the fact that any *analytic function of the complex variable* $x + iy$ satisfies the two-dimensional Laplace equation. Here, of course, i stands for the square root of -1, but we may note that, whatever value were assigned to i, we could write down the equation

$$\partial f(x + iy)/\partial y = if'(x + iy) = i\partial f(x + iy)/\partial x \tag{354}$$

for any differentiable function f and, if it were twice differentiable, go on to obtain

$$\partial^2 f(x + iy)/\partial y^2 = i^2 f''(x + iy) = i^2 \partial^2 f(x + iy)/\partial x^2. \tag{355}$$

Thus, if $i^2 = -1$, the expression $f(x + iy)$ satisfies eqn (346).

A function $f(x + iy)$ is said to be analytic in a region of the (x, y) plane if it is differentiable at all points of the region. Texts on functions of a complex variable prove that, in this case, it is *also twice differentiable* (and, indeed, differentiable any number of times) throughout the region.

Although these arguments identify $f(x + iy)$ as a solution of the two-dimensional Laplace equation, it is a complex number rather than a real quantity and therefore cannot itself represent a velocity potential. If, however, we put

$$f(x + iy) = \phi + i\psi, \tag{356}$$

then the real and imaginary parts of the equation

$$\partial^2(\phi + i\psi)/\partial x^2 + \partial^2(\phi + i\psi)/\partial y^2 = 0 \tag{357}$$

give us two separate pieces of information: that ϕ satisfies eqn (346); and that it is closely associated with another function ψ satisfying the same equation. Similarly, the real and imaginary parts of eqn (354) link ϕ and ψ through the two results

$$\partial\phi/\partial y = -\partial\psi/\partial x, \quad \partial\psi/\partial y = \partial\phi/\partial x. \tag{358}$$

These results are known as the Cauchy–Riemann equations, which are satisfied by the real and imaginary parts of any analytic function.

Not only does the real part ϕ of an analytic function $f(x + iy)$ satisfy eqn (346), but the converse result also holds. In fact, given any solution ϕ of eqn

(346), we can find a function ψ which satisfies eqns (358) so that ϕ is the real part[†] of an analytic function $f(x+iy) = \phi + i\psi$. For the existence of a function ψ satisfying

$$\partial\psi/\partial x = a(x, y), \quad \partial\psi/\partial y = b(x, y) \tag{359}$$

is assured if the integrability condition

$$\partial a/\partial y = \partial b/\partial x \tag{360}$$

is satisfied, and this integrability condition (eqn (360)) for the Cauchy–Riemann equations (eqns (358)) is simply the two-dimensional Laplace equation (eqn (346)).

We conclude that the theory of the potentials ϕ of two-dimensional irrotational flows is mathematically the same as the theory of analytic functions f of the complex variable $x + iy$. Furthermore, our conclusions automatically associate ϕ (the real part of f) with another function ψ (the imaginary part of f) which also turns out to be physically important.

The function ψ has the property, in fact, that *it takes a constant value along any streamline*. Thus, in terms of the components $(u, v, 0)$ of the two-dimensional velocity field, eqns (358) can be written

$$\partial\psi/\partial x = -v, \quad \partial\psi/\partial y = +u. \tag{361}$$

Accordingly, any displacement (dx, dy) along some curve on which the change

$$d\psi = (\partial\psi/\partial x)\,dx + (\partial\psi/\partial y)\,dy \tag{362}$$

is zero satisfies the condition

$$dy/dx = v/u. \tag{363}$$

This, of course, means that such a displacement along a curve $\psi = $ constant is in the direction of the local flow velocity.

An even simpler way of deriving the same conclusion begins by considering any smooth curve C on which we use the notations $\partial/\partial s$ and $\partial/\partial n$ to indicate differentiation along the tangent and along the normal at a point P on C, with the tangential direction obtained (Fig. 61) by rotation through $\frac{1}{2}\pi$ in the *positive* sense from the normal direction. Then we can write

$$\partial\phi/\partial n = \partial\psi/\partial s, \tag{364}$$

since we are free, locally, to choose axes with the x direction along the normal and the y direction along the tangent, in which case the second of the Cauchy–Riemann equations (eqns (358)) takes this form. Now, a streamline is such a curve C on which the normal velocity $\partial\phi/\partial n$ is zero at all points P along

---

[†] The statement that any solution of eqn (346) must be the real part of some analytic function is not, of course, in conflict with the fact that the imaginary part of f also satisfies eqn (346), since the imaginary part of f is necessarily the real part of the analytic function $-if(x+iy)$.

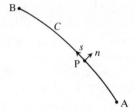

Fig. 61.   Illustrating the directions for differentiation along the tangent ($\partial/\partial s$) and along the normal ($\partial/\partial n$) at any point P on a smooth curve C.

the curve and it is therefore a curve with $\partial\psi/\partial s = 0$ so that the function ψ must take a constant value all along the streamline.[†]

The term *stream function*, used for ψ, recognizes this important property, and another which is an extension of it. Taking C now as a curve which is *not* necessarily a streamline and which joins two points A and B (Fig. 61), we can write the difference between the values of ψ at B and at A as an integral along C,

$$\psi_B - \psi_A = \int_C (\partial\psi/\partial s)\,\mathrm{d}s = \int_C (\partial\phi/\partial n)\,\mathrm{d}s, \tag{365}$$

which (since $\partial\phi/\partial n$ is normal velocity) represents the rate of volume flow streaming across C per unit span (that is, per unit length in the z direction). For an incompressible fluid, of course, this volume flow would be expected to take the same value for any curve with the same end points as C; which is why it can be written in a form ($\psi_B - \psi_A$) which depends only on those end points A and B, being the difference in the values of the stream function at B and at A.

For two-dimensional irrotational flows, then, we are led to make use of an analytic funtion $f(x + iy)$ whose real and imaginary parts both have major physical significance; as the velocity potential and the stream function respectively. The rather natural name given to this analytic function is the *complex potential*. We shall write it

$$\phi + i\psi = f(z), \text{ where } z = x + iy; \tag{366}$$

using the standard notation z for the complex variable because the danger of confusion with the use of z as the third Cartesian coordinate is insignificant in the context of this book. In fact, *only* the Cartesian coordinates x and y (and never z) are used in those parts of the book (Chapter 10 as well as the remainder of this chapter) which develop the two-dimensional theory. Conversely, the return of z as the third Cartesian coordinate in Chapter 11, which embeds

---

[†] The fact that the curves ψ = constant (the streamlines) everywhere intersect the curves ϕ = constant (the equipotentials) at right angles is, as this argument shows, a property shared by the imaginary and real parts of any analytic function.

certain results from two-dimensional theory in a three-dimensional framework, need cause no confusion because the results used have already been derived in previous chapters (so that there is no requirement to reintroduce the complex variable z at that point).[†]

For exactly similar reasons we here use the letter w for the *complex velocity* $f'(z)$, which was given by eqn (354) as $\partial(\phi + i\psi)/\partial x$. This expression allows us to identify its real and imaginary parts, using eqn (361), in the equation

$$w = f'(z) = u - iv. \tag{367}$$

In words, the derivative of the complex potential is an analytic function (the complex velocity) whose real and imaginary parts are the x component and *minus* the y component of the fluid velocity. We may note that the Cauchy–Riemann equations (eqns (358)) for *these* real and imaginary parts are the equations

$$\partial u/\partial y = \partial v/\partial x \text{ and } \partial u/\partial x + \partial v/\partial y = 0 \tag{368}$$

of irrotationality and continuity, respectively.

Briefly, we may also mention that *the stream function* (although not the velocity potential) *exists also for rotational flows* that are two-dimensional, so that the second of eqns (368) holds but not the first. In such a motion, the stream function must be defined through the eqns (361), which by themselves guarantee its characteristic properties; and this case of the general eqns (359) does satisfy the necessary integrability condition (eqn (360)) provided that the equation of continuity (second of eqns (368)) holds. For a two-dimensional velocity field $(u(x, y), v(x, y), 0)$ it is only the z component of vorticity

$$\zeta = \partial v/\partial x - \partial u/\partial y \tag{369}$$

that can be non-zero, and eqns (361) show that its relation to ψ takes the form

$$\partial^2\psi/\partial x^2 + \partial^2\psi/\partial y^2 = -\zeta. \tag{370}$$

This is a *modified form* of the two-dimensional Laplace equation which is quite frequently utilized for certain flows with non-zero vorticity.

Returning to the subject of *irrotational* two-dimensional flows, we study now the forms taken by the complex potential in both of the flows with *singularities* considered at the end of Section 9.1. A remarkably simple relationship between these two forms exists.

It is convenient to use polar coordinates r, θ which are related to x, y and the complex variable z by the equations

$$x = r\cos\theta, \quad y = r\sin\theta, \quad z = re^{i\theta}. \tag{371}$$

[†] Those few authors who have needed to make simultaneous use of the third Cartesian coordinate *and* the complex variable have commonly retained the letter z for both but used different type-faces.

For the *line source* which generates a radially outward velocity of magnitude given by eqn (352), the velocity components u, v are

$$u = j\cos\theta/(2\pi r), \quad v = j\sin\theta/(2\pi r), \tag{372}$$

so that the complex velocity (eqn (367)) takes the form

$$f'(z) = w = u - iv = j/(2\pi re^{i\theta}) = j/(2\pi z). \tag{373}$$

For the *line vortex*, which generates a velocity of magnitude $K/(2\pi r)$ at right angles to the radius, the velocity components are

$$u = -K\sin\theta/(2\pi r), \quad v = K\cos\theta/(2\pi r), \tag{374}$$

so that the corresponding complex velocity is

$$f'(z) = w = u - iv = -iK/(2\pi z), \tag{375}$$

which is the same as for the source but with j replaced by $(-iK)$.

In order to integrate either eqn (373) or eqn (375) to obtain the complex potential $f(z)$, we must use the properties of the *logarithm of a complex variable*

$$\log z = \int_1^z z^{-1}\,dz. \tag{376}$$

In particular, we use the property that $\log z$ *is necessarily many-valued* because the value of the integral (eqn (376)) depends on the path of integration followed and, in particular, on how many times that path encircles the origin. This is connected with the fact that the origin $z = 0$ is a *pole* of the integrand z^{-1} so that the integral increases by an amount $2\pi i$ (since the *residue* at that pole is 1) every time the origin is encircled once in the positive sense. The different possible values of eqn (376) all differ, therefore, by integer multiples of $2\pi i$.

Thus, when z is given by eqns (371) and the path of integration in eqn (376) is taken from 1 to r and then along the circle $|z| = r$ in the positive sense from r to z, we obtain

$$\log z = \left(\int_1^r + \int_r^z\right) z^{-1}\,dz = \log r + \int_0^\theta (re^{i\theta})^{-1}\, rie^{i\theta}\,d\theta = \log r + i\theta; \tag{377}$$

but, by including in the path n extra loops round the origin in the positive sense (where n may be positive or negative), the integral's value can be altered by $2n\pi i$. All this might be expected, of course, since the eqn (377) suggested by formally taking the logarithm in eqns (371) involves the polar angle θ which is only well defined by eqns (371) to within an arbitrary integer multiple of 2π.

For both the line source and the line vortex, with complex velocities given by eqns (373) and (375) respectively, the mathematical property that the integral (eqn (376)) is many-valued corresponds to a real characteristic of the flow. For the line source, we have

$$\phi + i\psi = f(z) = (j/2\pi)\log z = (j/2\pi)(\log r + i\theta); \tag{378}$$

thus, while the velocity potential ϕ simply takes the one value in eqn (353), the stream function ψ is given as

$$\psi = j\theta/(2\pi), \tag{379}$$

which is only well defined to within an arbitrary multiple of j. Equation (379) for ψ does, of course, correctly specify the streamlines as radii $\theta = $ constant; and it furthermore makes the rate of volume flow per unit span (from eqn (365)) across the curve C joining the two points A and B equal to a fraction $(\theta_B - \theta_A)/(2\pi)$ of the *total volume flow j* produced at the source O over a range 2π of angles, where $\theta_B - \theta_A$ is the angle subtended by A and B at O. On the other hand, the curve C could encircle O any number of times, positive or negative, requiring the addition of an arbitrary integer multiple of j to the volume flow across it; which clearly explains the many-valued character of the stream function (eqn (379)).

Similarly, for the line vortex, the complex potential is the many-valued function

$$\phi + i\psi = f(z) = (-iK/2\pi)\log z = (-iK/2\pi)(\log r + i\theta). \tag{380}$$

Thus, while the stream function is one-valued in this case, the *real* part of eqn (380) puts the velocity potential ϕ equal to its 'candidate value' in eqn (348). This agrees with the previous conclusion that, in the flow around a line vortex, there is no one-valued function ϕ whose gradient is the velocity vector, but that nevertheless the many-valued function (eqn (348)) which increases by K (the *circulation* around the vortex) every time the vortex is encircled in the positive sense does have its gradient equal to **u**.

We see that use of the complex potential $f(z)$ uncovers a certain complementarity in these flows: one has its stream function many-valued; the other its velocity potential. Furthermore, the mathematical fact that the ratio of the two eqns (378) and (380) is a 'pure imaginary' constant expresses the physical fact that the streamlines of one are the equipotentials of the other.

9.3 The method of conformal mapping

We conclude from the theory of Section 9.2 that two-dimensional irrotational flows can be evaluated if it is possible to determine a complex potential $f(z)$ which is an analytic function throughout a given domain D (the two-dimensional region occupied by fluid) and which, in addition, satisfies any necessary *limiting conditions* such as may apply either at solid boundaries or, perhaps, at very large distances from those boundaries. For example, in the model (Section 9.1) where a uniform stream of velocity

$$(U, V, 0) \tag{381}$$

is considered to be disturbed by a solid shape such as an aerofoil section being held at rest within it, the domain D is taken to be the region outside that solid shape and the complex velocity (eqn (367)) assumes the limiting value

$$f'(z) \to U - iV \tag{382}$$

for large values of $|z|$. Also, the condition $\partial\phi/\partial n = 0$ on the stationary solid boundary ∂D implies by eqn (364) that *the imaginary part of $f(z)$* (that is, the stream function ψ) *takes a constant value* on ∂D (where $\partial\psi/\partial s = 0$) or, more strictly, that the limiting value[†] which ψ takes on ∂D is constant. Again, near some point where a line source or line vortex is located, eqn (373) or eqn (375) represents the limiting form of $f(z)$ as that point is approached.

Just one extremely fruitful technique is applicable to all of these problems concerned with determining a function $f(z)$ that is analytic in a certain domain D and satisfies appropriate limiting conditions. This is the technique based on *conformal mapping*.

Two domains D and E are said to be conformally mapped onto one another if there exists a function $Z(z)$ which is an analytic function in D possessing these three properties:

(i)   there are no two points in D where the values taken by $Z(z)$ are the same;

(ii)   the values taken by $Z(z)$, of which we know from (i) that each is taken just once only, completely fill the domain E; and

(iii)   there is no point in D where the derivative $Z'(z)$ takes the value zero. As a matter of fact, it is unnecessary to specify property (iii) but, so as to avoid a break in the argument at this point, we postpone to the end of Section 9.3 a demonstration of the useful and interesting fact that (iii) is satisfied whenever (i) is satisfied.

Conformal mapping is a completely reciprocal relationship between two domains. Thus, properties (i) and (ii) imply that there is just one and only one point z in D for which $Z(z)$ takes a given value Z in E, and it is natural to use the simple notation $z(Z)$ for the inverse function specifying the point z in D that corresponds to the given point Z. With this notation, the function $z(Z)$ *is necessarily an analytic function in E* (that is, a differentiable function); for property (iii) specifies a non-zero limit $Z'(z)$ for the quantity dZ/dz which is obtained when we divide the difference between two neighbouring values of Z by the difference between corresponding values of z. Its reciprocal

$$(dZ/dz)^{-1} = dz/dZ \tag{383}$$

must, therefore, also tend to a non-zero limit $[Z'(z)]^{-1}$ which we can write as the derivative $z'(Z)$. Thus, the analytic function $z(Z)$ conformally maps E onto D.

[†] It is important to note that the definition of a domain D requires it to be a connected region which is also an *open* set, so that points on the boundary ∂D, while representing limits of points within D, are not themselves in D.

In the above argument it should be noted that the derivative $Z'(z)$ of the mapping function is a *complex* number, which can always be written as

$$dZ/dz = ke^{i\kappa},\tag{384}$$

with $k > 0$. Given a point z in D which corresponds to Z in E, and given any other point $z + dz$ situated within a very small circular neighbourhood of radius ε around z, we can write

$$dz = re^{i\theta} \text{ and } dZ = kre^{i(\theta+\kappa)} \text{ for } 0 < r < \varepsilon,\tag{385}$$

so that, in turn, the points dZ fill a very small circular neighbourhood of radius $k\varepsilon$ around Z, with their distances from the centre all multiplied by k and their angular positions rotated through an angle κ (Fig. 62). Locally, then, a conformal mapping takes the form of a simple magnification[†] and rotation

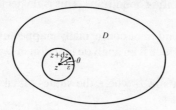

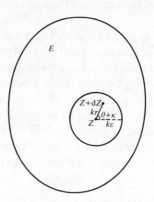

Fig. 62. In conformal mapping of D onto E, specified by an analytic function $Z(z)$ in D, a circle of small radius ε in D is mapped onto a circle of radius $k\varepsilon$ in E rotated through an angle κ in the positive sense, if $ke^{i\kappa}$ is the value of $Z'(z)$ at the centre of the circle.

[†] We refer, of course, to 'magnification' by a factor k even in the case $k < 1$ when a shrinking occurs.

without any deformation or *straining* (Section 4.3). In particular, two radii with $\theta = \theta_1$ and θ_2 in D, which intersect at an angle $\theta_2 - \theta_1$, correspond to two radii in E which intersect at the angle

$$(\theta_2 + \kappa) - (\theta_1 + \kappa) = \theta_2 - \theta_1. \tag{386}$$

This shows that conformal mapping *cannot alter the angle* at which two lines intersect at an interior point.

An earlier footnote pointed out, incidentally, that a domain D is an open set so that every point z in D is such an interior point possessing a small circular neighbourhood lying entirely within D. By contrast, the boundary ∂D consists of limit points of D which are not themselves in D. The mapping must evidently associate these with limit points of E which are not themselves in E (in short, with ∂E), but we shall see that not *all* of the general results proved above for D and E are necessarily valid for ∂D and ∂E.

The conformal mapping of two domains D and E onto one another has proved to be a powerful method for analysing two-dimensional irrotational flows. Given, for example, a complex potential $f(z)$ in D (that is, an analytic function satisfying appropriate limiting conditions), we can immediately write down another complex potential satisfying corresponding limiting conditions in E, namely the function

$$f(z(Z)). \tag{387}$$

For this is an analytic function of Z since the limit of

$$df/dZ = (df/dz)(dz/dZ) \tag{388}$$

necessarily exists (as the product of the limits of the two quantities in brackets, which both exist since $f(z)$ and $z(Z)$ are analytic functions). Furthermore, if the imaginary part of f takes constant limiting values on ∂D it must take the same constant limiting values on the corresponding boundary ∂E. In addition, we can ensure that the limit (382) of df/dz for large values of $|z|$ is the same as that of df/dZ simply by using a mapping such that in this limit

$$dz/dZ \to 1. \tag{389}$$

A first indication of the fruitfulness of this method is given in Section 9.4, where a single mapping function $Z(z)$ is used to derive many interesting flows; we may add that in Chapter 10 very many more flows (including some with circulation) are obtained from this same first choice of $Z(z)$. Before thus exemplifying the method, however, we conclude Section 9.3 by demonstrating that the property (iii) of any conformal mapping (that $Z'(z) \neq 0$ in D) is a *necessary* consequence of the property (i) (that $Z(z)$ takes no value twice in D). We do this by showing that, if there were a point z_0 in D where $Z'(z_0) = 0$, then for any point z_1 *sufficiently close* to z_0 there must be at least one other value of z with

$$Z(z) = Z(z_1). \tag{390}$$

Suppose, in fact, that n is the smallest integer for which the nth derivative $Z^{(n)}(z_0)$ is non-zero. We then prove that the eqn (390) has n solutions for z (including the solution $z = z_1$) within a small neighbourhood of $z = z_0$. Thus the hypothesis that $n > 1$ is in contradiction with the assumption that $Z(z)$ takes no value twice.

The idea of the proof is that the definition of n makes the Taylor series expansion for $Z(z)$ about $z = z_0$ take the form

$$Z(z) - Z(z_0) = (z - z_0)^n Z^{(n)}(z_0)/n! + O\left(|z - z_0|^{n+1}\right), \qquad (391)$$

so that, at distances from z_0 which are sufficiently small for the error term in eqn (391) to be neglected, eqn (390) becomes

$$(z - z_0)^n Z^{(n)}(z_0)/n! = (z_1 - z_0)^n Z^{(n)}(z_0)/n!, \qquad (392)$$

which has n roots, given by

$$(z - z_0) = (z_1 - z_0)e^{2\pi i m/n} \qquad (393)$$

for $0 \leqslant m < n$ (Fig. 63). Here, however, in order to guard against the possibility that eqn (390) is satisfied only approximately rather than exactly at values of z other than z_1, we can use Cauchy's theorem: the integral

$$\frac{1}{2\pi i} \int_C \frac{Z'(z)\,dz}{Z(z) - Z(z_1)} \qquad (394)$$

around a small circle C of centre z_0 and radius 2ε (where $|z_1 - z_0| = \varepsilon$) is equal to the sum of the residues of the integrand at its poles, and these are the

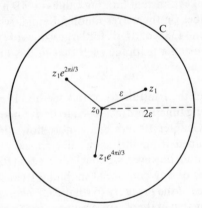

Fig. 63. If there were a point z_0 where the lowest-order non-vanishing derivative of $Z(z)$ was (say) the third, $Z'''(z_0)$, then for a point z_1 with small enough $\varepsilon = |z_1 - z_0|$ the function $Z(z)$ would take the value $Z(z_1)$ three times, as we show by estimating the integral (expression (394)) taken around a circle C of centre z_0 and radius 2ε.

solutions of eqn (390) within C, at each of which the residue is 1. Thus, the integral (expression (394)) represents the exact *number of solutions* of eqn (390) within C.

Now, for ε small enough, we can approximate $Z(z)$ in both the numerator and denominator of expression (394) by the leading term in eqn (391) with a proportional error as small as we please, and can thus infer that the integral (expression (394)) takes the value n (the number of solutions of eqn (392)), with a proportional error as small as we please. Since, however, the integral (expression (394)) *is necessarily an integer* (the number of solutions of eqn (390)) we deduce that for ε small enough it must be exactly n. This completes the proof that properties (i) and (ii) by themselves are sufficient to specify a conformal mapping, for which property (iii) necessarily follows from (i).

9.4 Conformal mapping exemplified

A single mapping function

$$Z(z) = z + a^2 z^{-1}, \tag{395}$$

where $a > 0$, is used in this section and in Chapter 10 to demonstrate the fruitfulness of the method of conformal mapping while giving a general introduction to aerofoil theory. Readers may also find the present comprehensive account of how eqn (395) can be used a helpful summary of techniques for deriving as much value as possible from other mapping functions, whether they have been obtained by the classical analytical methods described in the general literature or by computational procedures.

In the polar coordinates (eqn (371)) for z, the function $Z(z)$ takes the form

$$Z = re^{i\theta} + a^2 r^{-1} e^{-i\theta}, \tag{396}$$

which may be written

$$Z = X + iY \text{ with } X = (r + a^2 r^{-1}) \cos \theta, Y = (r - a^2 r^{-1}) \sin \theta. \tag{397}$$

It follows that a circle $r = c$, on which

$$x = c \cos \theta, \quad y = c \sin \theta, \quad z = ce^{i\theta}, \tag{398}$$

becomes a curve

$$X = A \cos \theta, Y = B \sin \theta, \text{ with } A = c + a^2 c^{-1}, B = c - a^2 c^{-1}. \tag{399}$$

This curve is like the original circle (eqns (398)) except that it is *elongated* in the X direction (since $A > c$) and *foreshortened* in the Y direction (since $B < c$). Thus, it is an ellipse with semi-axes A and B, provided that $c > a$ (so that $B > 0$).

We may note that, just as the coordinates (eqns (398)) for $-\pi < \theta \leqslant \pi$ represent all points on the circle just once, so the coordinates (eqns (399)) for $-\pi < \theta \leqslant \pi$ represent all points on the ellipse just once. Also, the semi-axis A

is an increasing function of c for $c > a$, taking all values greater than $2a$ just once; while the semi-axis B, too, is an increasing function of c, taking all values greater than zero just once.

It follows that the analytic function (eqn (395)) gives a conformal mapping from the domain D defined as *the outside of the circle $r = a$* (so that the points (eqns (398)) with $-\pi < \theta \leqslant \pi$ and $c > a$ fill it just once) onto a domain E where the points (in eqns (399)) with $-\pi < \theta \leqslant \pi$ and $c > a$ represent a set of ellipses filling it just once; with the semi-axes A and B taking all values greater than $2a$ and all values greater than 0, respectively (Fig. 64). We see that E includes all points in the plane except those on the strip

$$-2a \leqslant X \leqslant 2a, \quad Y = 0, \tag{400}$$

which represents the limiting case of the ellipse (eqns (399)) as $c \to a$. This strip (expression (400)) is excluded from E by the requirement that $c > a$. Thus, we can define E as *the outside of the strip* (expression (400)).

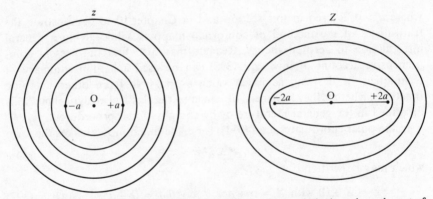

Fig. 64. The special conformal mapping (eqn (395)), used extensively throughout the rest of Chapter 9 and the whole of Chapter 10, maps the outside of the circle of centre the origin and radius a in the z plane onto the outside of the strip given by expression (400), which extends from $Z = -2a$ to $Z = 2a$, in the Z plane. Any circle of larger radius $c > a$ in the z plane is mapped onto the ellipse given by expression (399) in the Z plane.

The mapping function (eqn (395)) gives an excellent illustration of the principle that properties known to hold in the conformal mapping between two domains D and E cannot necessarily be assumed to hold on their boundaries ∂D and ∂E. For example, property (iii) that $Z'(z)$ is non-zero throughout D cannot be extended to ∂D since

$$Z'(z) = 1 - a^2 z^{-2} \tag{401}$$

vanishes at the two points $z = \pm a$ of the boundary ∂D, corresponding to two points $Z = \pm 2a$ of the boundary ∂E. Again, the property (eqn (386)) of conformal mapping that it preserves angles between radii which meet at any internal point is not satisfied at these two boundary points; thus, the range of angles of radii approaching one of them from within the domain is π in D but 2π in E (Fig. 64). In fact, if z_0 represents either of the points where $Z'(z_0) = 0$ the mapping function locally takes the form of eqn (391) with $n = 2$, where the term in $(z - z_0)^2$ produces such a doubling of angles.

Furthermore, we may observe that on the strip given by expression (400) which corresponds to the circle $r = a$, the values of X and Y given by eqns (397) are

$$X = 2a\cos\theta, \quad Y = 0, \tag{402}$$

and that these values are unchanged when θ is replaced by $-\theta$. Thus, we are at first drawn to conclude that two different points $\pm\theta$ on $r = a$ are mapped on to the same point (eqns (402)); and, indeed, to relate this conclusion to the breakdown of property (iii) on the boundary. Nevertheless, although this conclusion is correct in its most literal sense, there are some good mathematical reasons and some excellent physical reasons for formulating it rather differently.

Physically, the domain E (the outside of the strip given by expression (400)) is used to model the interaction between *a thin solid plate* (whose thickness is taken as effectively zero) and a fluid moving outside it. Now, the eqns (397) for a point in E represent, when r is only *just* greater than a, a fluid particle just *above* such a plate when $\sin\theta$ is positive ($0 < \theta < \pi$), and a fluid particle just *below* the plate when $\sin\theta$ is negative ($-\pi < \theta < 0$). Evidently, then, as $r \to a$, the point in E tends to a point on the *upper* surface of the plate when $\sin\theta > 0$, and to a point on the *lower* surface when $\sin\theta < 0$. Now, the actual fluid at such points on the upper and lower surfaces of the plate is completely separate; for example, the plate can support a difference of *pressure* at the two points. Thus, we can effectively claim that the mapping function (eqn (395)) maps points on the boundary $r = a$ of D onto different points in E: on the upper surface of the plate if $0 < \theta < \pi$, and on the lower surface if $-\pi < \theta < 0$.

Independently of such physical considerations, mathematicians studying the theory of functions of a complex variable found long ago that certain distinctions needed to be drawn along very similar lines. They began to refer to a domain such as E (the whole complex plane except for the strip given by expression (400)) as being a *cut plane*, imagining it as a piece of paper (representing the complex plane) that had been cut from $Z = -2a$ to $Z = 2a$. Then they viewed any limiting process, in which a point in D with $r > a$ tended to the boundary $r = a$, as being mapped into a process in E as Z tended to the *upper edge* of the cut if $0 < \theta < \pi$, and to the *lower edge* of the cut if $-\pi < \theta < 0$. Some such nomenclature was seen as necessary because functions (including $z(Z)$ itself) that are analytic in E are able to have different limiting values as Z tends to a point on the strip given by expression (400) from above and below. It was therefore convenient to refer to such limiting values as values on the upper and lower edges of the cut, respectively.

Henceforth we use the physically motivated terminology of fluid flowing around a flat plate whose upper and lower surfaces are mapped on to the upper and lower halves of the circle $r = a$. However, any reader who prefers it is of course free to regard those upper and lower surfaces as being the upper and lower edges of a cut in the complex Z plane.

The general method for using the mapping function (eqn (395)) to derive two-dimensional irrotational flows is to move backwards and forwards between the domains D and E. The method depends upon being prepared:

(i)   to use any known flow in either D or E to deduce the corresponding flow in the other domain; and

(ii)   to proceed similarly with regard to any *piece* of D (that is, to any domain D_1 contained within D which the mapping function maps onto a domain E_1 contained within E).

Giving first of all a very simple illustration of (i), we note that in domain E the flat plate does not disturb at all a uniform flow with velocity $(U, 0, 0)$ parallel to it, and therefore with complex velocity U and complex potential

$$f(Z) = UZ. \tag{403}$$

Applying the mapping eqn (395) we obtain the corresponding flow in domain D as

$$f(z) = U(z + a^2 z^{-1}). \tag{404}$$

This represents the irrotational motion produced when a circular cylinder of radius a disturbs the same uniform stream. The complex velocity at large distances from the body is seen to be unchanged, as must always be so for any mapping such as eqn (395) which satisfies the limiting condition (expression (389)).

The conclusion (eqn (404)) gives a link with the methods of Chapter 8, being in agreement with the conclusion that a continuous distribution of dipoles of strength given by eqn (329) per unit length along the axis of the cylinder (that is, a *line dipole*, with complex potential $U a^2 z^{-1}$) represents its interaction with a uniform stream at right angles to that axis. By contrast, most results obtained later by the method of conformal mapping are not easily obtained by other methods. In the meantime, the complex potential already possesses one clear advantage: that those lines (Fig. 65) on which the imaginary part of f is constant are streamlines.

Needless to say, the irrotational motion (eqn (404)) cannot be realized as a steady flow. Its distribution of complex velocity

$$w = f'(z) = U(1 - a^2 z^{-2}) \tag{405}$$

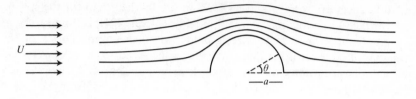

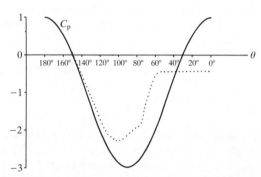

Fig. 65. Streamlines in the irrotational motion (eqn (404)) around a circular cylinder. IMPORTANT NOTE: ALL STREAMLINE PATTERNS FOR TWO-DIMENSIONAL FLOWS GIVEN IN THIS BOOK ARE PLOTTED AT EQUAL INTERVALS OF THE STREAM FUNCTION ψ; THE DISTANCE BETWEEN ANY TWO ADJACENT STREAMLINES IS THEN A VALUABLE INDICATION OF THE AVERAGE FLUID SPEED BETWEEN THEM (TO WHICH, INDEED, IT IS INVERSELY PROPORTIONAL). The lower diagram plots against the polar coordinate θ the pressure coefficient (eqn (408)) which, in steady flow, would be associated with this irrotational motion (plain line), together with an experimentally observed distribution of the surface pressure coefficient in a high-Reynolds-number case when the boundary layer becomes turbulent before separation.

would on the cylinder surface take the value

$$w = u - iv = (2U \sin \theta)ie^{-i\theta} \text{ on } z = ae^{i\theta}, \tag{406}$$

representing a fluid speed

$$q = 2U |\sin \theta| \tag{407}$$

directed at right angles to the radius. The surface flow would accelerate from a stagnation point at $\theta = \pi$ to a maximum speed of $2U$ (twice the speed of the undisturbed stream) at $\theta = \frac{1}{2}\pi$ before undergoing a massive retardation which must not only bring about separation of the boundary layer (Section 5.6) but also produce modifications to the flow on an even more extensive scale than in the case of the sphere (Fig. 49). Figure 65 indeed compares the distribution of pressure coefficient observed in steady flow at Reynolds numbers large enough for the boundary layer to become turbulent before separation (indicated, as in Fig. 49, by a dotted line), with the irrotational-flow value

$$C_p = 1 - 4 \sin^2 \theta. \tag{408}$$

The differences are seen to be very much more marked than for the sphere.

Nevertheless, the theoretical irrotational flow (eqn (404)) is found to be of very great importance as a starting-point for further applications of the technique of conformal mapping, many of which lead to flows of very real practical relevance. Some of these, to be described in Chapter 10, involve new utilizations of the idea labelled (i) on p. 168. These either work from the flow of Fig. 65 *rotated through an angle* (which must be an equally valid complex potential in D, and which is mapped into a non-trivial flow in E), or from the flow of Fig. 39, or from a combination of both. In the rest of this chapter, however, we concentrate on illustrating the idea labelled (ii) on p. 168.

Now, the fact that eqn (404) represents the irrotational flow in D around a circle of radius a implies that the corresponding flow around a circle of (say) radius $c > a$ must take the same form with c replacing a, namely

$$f(z) = U(z + c^2 z^{-1}).\qquad(409)$$

This is a flow in a domain D_1 contained within D and consisting of the outside of the circle $r = c$. The mapping function (eqn (395)) maps D_1 onto a domain E_1 contained within E, namely the outside of the ellipse given by eqns (399). Therefore it must map the flow (eqn (409)) onto the corresponding flow around this ellipse. By analogy with the case of the Rankine ovoid (Fig. 48), we can expect that, when the ratio B/A of the ellipse's semi-axes is fairly small, the modifications to the irrotational flow due to boundary-layer separation may be only modest.

The complex velocity in E_1 takes the form of eqn (388), which with the help of eqn (398) can be written

$$w = f'(Z) = U(1 - c^2 z^{-2})(1 - a^2 z^{-2})^{-1}.\qquad(410)$$

On the ellipse given by eqns (399), which corresponds to the circle given by eqns (398), the eqn (410) becomes

$$w = u - iv = U(1 - e^{-2i\theta})(1 - a^2 c^{-2} e^{-2i\theta})^{-1},\qquad(411)$$

where eqns (399) express $a^2 c^{-2}$ as

$$a^2 c^{-2} = (A - B)/(A + B)\qquad(412)$$

in terms of the semi-axes A and B; so that the fluid speed takes the form

$$q = |w| = U(A + B)|\sin\theta|(A^2 \sin^2\theta + B^2 \cos^2\theta)^{-\frac{1}{2}}.\qquad(413)$$

Figure 66 shows how the distribution (eqn (413)) of fluid speed varies as a function of the X coordinate on the ellipse given by eqns (399), for a small value (0.15) of the semi-axis ratio B/A. The maximum value attained is $1.15\,U$ (in general, $U(A + B)/A$) and there is only a very weak retardation for a long distance beyond this maximum. Much closer to the rear stagnation point, however, the retardation becomes strong. In this region, boundary-layer

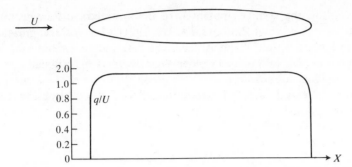

Fig. 66. The distribution (eqn (413)) of the fluid speed q as a function of the X coordinate of a point on the surface of an elliptic cylinder of axis-ratio 0.15 in irrotational motion with oncoming velocity U.

separation is observed as expected, but the influence of the resulting wake on the pressure distribution ahead of separation is only rather modest.

An aerofoil is a shape (Fig. 60) which, in symmetrical flow around it, avoids that rear stagnation point *and* the associated boundary-layer separation by possessing a sharp trailing edge. It retains, however, for several good reasons (see Chapter 10), a well-rounded leading edge similar to that of the ellipse of Fig. 66. It makes a transition, in short, from such elliptical shaping at the front to a property similar to that illustrated for the flat plate's rear end in Fig. 64, and there related to the vanishing of the derivative $Z'(z)$ of the mapping function at the point in question.

These considerations suggest that, in order to obtain an aerofoil shape, the same mapping function should be applied to a different domain D_2 contained within D. Here D_2 is the outside of a circle which at the *rear* end goes through the point $z = a$ where $Z'(z) = 0$ (in order to generate the sharp trailing edge) but which, at the front end, imitates the circle $r = c > a$ by going through the point $z = -c$ where $Z'(c) \neq 0$. We need, in short, a circle of intermediate radius

$$b = \tfrac{1}{2}(a+c),\tag{414}$$

centred on

$$z = -\delta \text{ with } \delta = \tfrac{1}{2}(c-a).\tag{415}$$

When it disturbs a uniform stream of velocity U, the complex potential in D_2 must be

$$f(z) = U[(z+\delta) + b^2(z+\delta)^{-1}];\tag{416}$$

essentially as in eqn (404) or eqn (409), but with the radius changed to b and the centre to $z = -\delta$. The mapping function (eqn (395)) must transform D_2 into a domain E_2 representing the outside of an aerofoil section, and also transform eqn (416) into the complex potential of the symmetrical flow around this aerofoil. It was Nikolai Zhukovski (1847–1921) who first recognized these facts, and the aerofoil shape is therefore referred to as the *symmetrical Zhukovski aerofoil* (although in older Western texts his Russian name was transliterated as Joukowsky).

Figure 67 shows the aerofoil shape obtained for $\delta/b = 0.1$ by using eqn (395) to map the points

$$z = -\delta + be^{i\sigma} \tag{417}$$

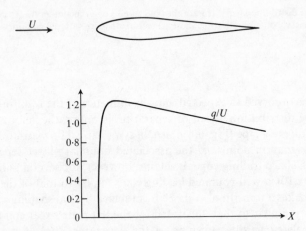

Fig. 67. The distribution (eqn. (420)) of the fluid speed q as a function of the X coordinate of a point on the surface of a symmetrical Zhukovski aerofoil in a stream of velocity U.

of ∂D_2 (where the polar coordinate relative to eqn (415) has been written σ to avoid confusion with θ). The corresponding coordinates (X, Y) of a point on the aerofoil surface are obtained from the real and imaginary parts of eqn (395) as

$$\left.\begin{array}{l} X = 2\,(b\cos\sigma - \delta)\,(b^2 - 2b\delta\cos\tfrac{1}{2}\sigma + \delta^2)\,(b^2 - 2b\delta\cos\sigma + \delta^2)^{-1}, \\[2mm] Y = 2b^2\delta\sin\sigma(1 - \cos\sigma)\,(b^2 - 2b\delta\cos\sigma + \delta^2)^{-1}. \end{array}\right\} \tag{418}$$

The complex velocity df/dZ is derived from eqn (416) and eqn (388) as

$$w = f'(Z) = U[1 - b^2(z + \delta)^{-2}]\,(1 - a^2 z^{-2})^{-1}, \tag{419}$$

and the distribution of fluid speed $q = |w|$ over the aerofoil surface is found to be

$$q = (U\cos\tfrac{1}{2}\sigma)\,(b^2 - 2b\delta\cos\sigma + \delta^2)b^{-1}\,[(b^2 - 2b\delta)\cos^2\tfrac{1}{2}\sigma + \delta^2]^{-\frac{1}{2}}. \tag{420}$$

Its plot against X in Fig. 67 fulfils our expectation that *a weak retardation is maintained* all the way to the trailing edge of the aerofoil in this case.

Experiments confirm that boundary-layer separation is indeed avoided on such shapes. Nevertheless, the chief importance of aerofoil sections lies in their behaviour, not in a symmetrical oncoming stream, but in an inclined stream. This behaviour represents the main subject matter of Chapter 10.

10
Flows with circulation

The topology of three-dimensional space is such that the externally un-bounded region outside a body of finite size (assumed without any holes through it) is simply connected, in the sense (Section 6.2) that for any closed curve C within the region there exists a surface S within the region which spans the closed curve C. For irrotational flow in such a region, a single-valued velocity potential (eqn (182)) necessarily exists. The flow, furthermore, is uniquely determined by the normal component $\mathbf{u}_s \cdot \mathbf{n}$ of the instantaneous velocity of the solid boundary. This result was proved in Section 6.2 for finite regions of fluid, and was extended at the end of Section 8.2 to externally unbounded fluid whose velocity tends to zero at large distances. The whole irrotational motion can be generated instantaneously by impulsively applying those boundary velocities, a process which requires a total impulse given by eqn (304), and a total kinetic energy given by eqn (310), to be imparted to the fluid. For a simple translational motion of the body with constant velocity, the impulse $\mathbf{I}$ is constant, implying eqn (335) which states that *all* components of the total force acting between the body and the fluid are zero. Earlier, the constancy of energy had been used to deduce eqn (240) which states that the drag (the component D of force opposing the body's motion) is zero—a result whose importance lies in the fact that streamlined bodies (Section 7.4), designed so that the flow will be irrotational outside a very thin boundary layer and wake, are subject to very *low drags* (drags related, as in eqn (345), to the defect of volume flow in this thin wake).

In *two dimensions*, however, the externally unbounded region outside a body section of finite size cannot be simply connected; because a closed curve C embracing that body section cannot be spanned by any surface S within the two-dimensional region (by contrast, a dome-like surface is always available in three dimensions to span such a curve). The region is, nevertheless, *doubly connected* in the sense (Section 6.4) that we can choose a particular closed curve C_1 which embraces the body section just once, and such that every closed curve in the region can be deformed, entirely within the region, into the closed curve C_1 taken n times (where n may be positive, negative, or zero). In such a region, an irrotational flow is uniquely determined if the circulation K around C_1 is specified *in addition* to the motions of the solid boundary. It is, however, only the flow with $K = 0$ which possesses a single-valued velocity potential and which can be generated instantaneously by actions at the boundary which impart an impulse $\mathbf{I}$ to the fluid. For the other flows, with $K \neq 0$, this is not so and we therefore cannot apply the argument given by eqn (335) to the effect that all components of the total force acting between the body and the fluid (assumed to be in steady relative motion) must be zero.

In this chapter, we analyse the flow occurring when an aerofoil section moves steadily through fluid at a positive, but not too large, angle of incidence, and we explain how, *not instantaneously* but over a definite period of time, a flow is set up which is close to an irrotational flow with a certain non-zero value of the circulation K (close, that is, except within a thin boundary layer and wake). We show, too, that in this irrotational flow the force with which the fluid acts on the body takes a rather substantial value L which is *directed at right angles* to the direction of relative motion. Such a force is of great engineering interest because it requires no continuously applied power (the scalar product of velocity and force) to maintain it. The real flow, of course, must exhibit, in addition to L, a modest drag force D opposing the motion, which is associated with the action of viscosity and turbulence in a thin boundary layer (and which is, as before, related to a defect of volume flow in the wake); some small expenditure of power is required in order to overcome this drag force D. Nevertheless, a crucial conclusion from the present analysis is that *an aerofoil has the capability of generating a force L at right angles* to the direction of its relative motion which is enormously greater than the modest drag force opposing that motion.

We shall refer to this force L on an aerofoil as 'lift', because its best known use occurs when the aerofoil is a section of an aircraft wing moving horizontally through air; then the force L is directed upwards, and constitutes the main mechanism for supporting the aircraft's weight. It is, indeed, usual to describe the force L directed at right angles to *any* aerofoil's motion as 'lift', even in those many interesting cases when L does *not* act vertically. Some of these are described in Chapter 11, which proceeds also to a second approximation, based on taking the two-dimensional flow as a first approximation locally while allowing for the three-dimensional character of the flow as a whole. In the meantime, however, the present chapter is exclusively confined to analysing how circulation and lift arise in a variety of purely two-dimensional motions.

10.1 Rotating cylinders

We begin by describing flow around rotating cylinders because it is clearer in *qualitative* terms that circulation might be generated by rotation (even though, as Section 10.2 shows, a more quantitative analysis becomes possible for aerofoils at incidence). For example, the flow illustrated in Fig. 39 is capable of being generated from rest, if enough time is allowed, by rotating both the inner cylinder and the outer cylinder in such a way that the boundary velocity of each takes the value $K/(2\pi r)$ appropriate to its radius. Initially, such rotation would generate equal amounts of vorticity of opposite signs in thin boundary layers attached to the cylinders. Equation (165) tells us, however, that diffusion with diffusivity v would cause both layers to grow initially like $(vt)^{\frac{1}{2}}$; and, after a time long enough for this distance to exceed significantly the separation between the cylinders, diffusion must have so mixed the vorticity of opposite signs as to reduce net vorticity to the level when the flow is effectively the irrotational flow, with circulation, of Fig. 39.

This rather obvious, but very slow, method of using cylinder rotation to generate a flow with circulation has been mentioned, however, only in order to emphasize that in the rest of this section we are concerned with much faster mechanisms. Their time for effective operation is related, not to the time needed for a *diffusion* distance to be comparable with the body dimensions, but to the much shorter time required for convection of vorticity by an external flow to carry it well clear of the body. Before describing these mechanisms, however, we consider from a purely mathematical standpoint the range of irrotational motions that are possible when a stationary circular cylinder of radius a disturbs a uniform stream of velocity U. These are important, of course, not only in their own right but because the technique of conformal mapping can be used (Section 10.2) to infer corresponding flows about aerofoils.

Mathematically, the theory of Section 6.4 leads us to expect (as noted earlier) that the flow must be uniquely determined if the circulation around the cylinder is specified. For if there were two flow fields u_1 and u_2 with the same circulation and the same velocity at large distances, their difference $u = u_1 - u_2$ would have zero circulation and so possess a single-valued velocity potential ϕ. Also, the velocity field u would tend to zero at large distances. Moreover, we do not need any special theory of its behaviour in the far field, such as was used for three-dimensional flows (Section 8.2) to extend uniqueness results to externally unbounded fluid. This is because ϕ is the real part of a complex potential $f(z)$ which is an analytic function (Section 9.2), and is subject therefore to Laurent's theorem stating that it can be expanded as a series

$$f(z) = f_0 + f_1 z^{-1} + f_2 z^{-2} + \ldots \tag{421}$$

for large $|z|$. In that limit, then, ϕ tends to a constant and the complex velocity $f'(z) = u - iv$ is of order $|z|^{-2}$, so that an integral of $(\phi u) \cdot n$ over a large circle $|z| = $ constant can be made as small as we please, thus allowing the 'uniqueness' conclusion that $u = 0$ (giving $u_1 = u_2$) to be derived precisely as at the end of Section 8.2.

Knowing that there is only one flow with circulation K which can result when the circular cylinder of radius a, referred to as ∂D in Section 9.4, disturbs a uniform stream with velocity $(U, 0, 0)$, we are free to use any simple method (such as the adding together of two known flows) in order to obtain a motion with these properties, which we can be sure is 'the' solution to the problem (a similar approach was used much earlier in obtaining Fig. 37). In fact, the two complex potentials we linearly combine in this way are those given by eqn (404) which represents the required flow in a case with zero circulation, and eqn (380) which represents a line vortex of strength K stretching along the axis of the cylinder. In the latter case, of course, we use only a piece of the flow, namely the piece within the domain D (that is, outside the cylinder). This piece satisfies the boundary condition on ∂D, where its imaginary part is constant, while its complex velocity tends to zero at large distances, so that it in no way alters the conditions on ∂D and at large distances is already satisfied by eqn (404).

Actually, we prefer to *subtract* eqn (380) *from*, rather than add it to, eqn (404) because it turns out that the most physically interesting flows have circulation K in the *negative* sense around the cylinder. We obtain

$$f(z) = U(z + a^2 z^{-1}) + (iK/2\pi) \log z. \tag{422}$$

The complex velocity is

$$w = f'(z) = U(1 - a^2 z^{-2}) + (iK/2\pi) z^{-1}, \tag{423}$$

which, as required, takes the limiting value U for large $|z|$. Its boundary value is

$$w = ie^{-i\theta}[2U \sin \theta + (K/2\pi a)] \quad \text{on} \quad z = ae^{i\theta}, \tag{424}$$

which represents a fluid speed

$$q = 2U|\sin \theta + (K/4\pi a U)| \tag{425}$$

directed at right angles to the radius.

Equation (425) tells us that the presence of circulation has *shifted the stagnation points* from their symmetrically placed positions $\theta = 0$ and $\theta = \pi$ (where eqn (407) vanishes) to new positions

$$\theta = -\beta \quad \text{and} \quad \theta = \pi + \beta, \tag{426}$$

where β is the acute angle such that

$$\sin \beta = (K/4\pi a U). \tag{427}$$

Both the new positions are below the x axis.

The stream function ψ is given by the imaginary part of eqn (422) as

$$\psi = U(r - a^2 r^{-1}) \sin \theta + (K/2\pi) \log r. \tag{428}$$

This allows us to plot the streamlines $\psi = $ constant, as is done in Fig. 68 for the case $\beta = 15°$. Of these, the streamline

$$\psi = (K/2\pi) \log a \tag{429}$$

is the *dividing streamline* which approaches the body surface $r = a$ at the left-hand stagnation point, where it divides to encircle that surface on both sides before leaving it at the right-hand stagnation point.

We may note that, at each stagnation point, the dividing streamline is directed *normally* to the body surface. Indeed, at a stagnation point z_s of any

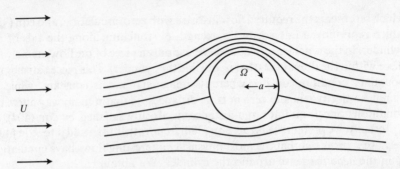

Fig. 68. Streamlines in the irrotational motion (eqn (422)) around a rotating circular cylinder of radius a, when the circulation K around the cylinder takes the value defined by eqn (427) with $\beta = 15°$.

flow, the complex velocity $f'(z_s)$ is zero. Then, provided that the second derivative $f''(z_s)$ is non-zero, the local behaviour of the complex potential is as

$$f(z) - f(z_s) = \tfrac{1}{2}(z - z_s)^2 f''(z_s) + 0(|z - z_s|^3), \tag{430}$$

and therefore *the directions of the streamlines* through $z = z_s$ (given by requiring the imaginary part of eqn (430) to vanish) *must be at right angles* (since if $\tfrac{1}{2}(z - z_s)^2 f''(z_s)$ is real it remains real when $z - z_s$ has been multiplied by i).

The streamlines are seen, furthermore, to be more bunched together above the cylinder than below, corresponding to the fact that the fluid speed (eqn (425)) takes greater values where $\sin \theta$ is positive ($0 < \theta < \pi$) than where it is negative ($-\pi < \theta < 0$). It follows that, in steady flow, the excess pressure

$$p_e = \tfrac{1}{2}\rho(U^2 - q^2) \tag{431}$$

takes greater values below the cylinder than above, tending to impart a *lift* to the cylinder. Defining the lift L as the force in the y direction (that is, perpendicular to the undisturbed stream) per unit span (that is, per unit length of cylinder), we have

$$L = -\int_{-\pi}^{\pi} (p_e \sin \theta)\, a\, d\theta \tag{432}$$

and, when p_e is substituted from eqns (431) and (425), it is only the term $(-\rho U K/\pi a) \sin \theta$ therein (and not the terms representing even functions of θ) that can contribute to eqn (432). We thus obtain for this case the simple connection

$$L = \rho U K \tag{433}$$

between circulation and lift, which is shown later (Section 10.3) to be a general property of flows with circulation. The drag, on the other hand, is zero for this irrotational flow, as may be calculated by replacing $\sin\theta$ with $\cos\theta$ in eqn (432), and which is in any case obvious from the left–right symmetry of Fig. 68.

It is not, of course, possible to claim that the theoretical flow pattern depicted in Fig. 68 represents any real flow to a close approximation, and one of its gravest deficiencies is, indeed, its failure to predict any drag. In certain other respects, however (including the prediction of lift), it gives a good general indication of the characteristics of flow around a circular cylinder which is rotated in the 'negative' (that is, clockwise) sense, and it is accordingly of real interest to consider the mechanism by which such rotation acts to generate circulation in this case.

For this purpose, it is necessary to recall eqn (160) which states that *the strength of a vortex sheet* is equal to the change in tangential velocity across the sheet. In the flow past the nonrotating cylinder represented by the dotted line in Fig. 66, the boundary layer separates at two points symmetrically placed above and below the cylinder. At each of these points, the boundary layer just before separation is a vortex sheet with the *same* strength, equal to the flow velocity V just outside that boundary layer. Thus, vorticity is shed into the wake at those points in amounts which are *equal but of opposite sign*, so that no net vorticity is shed into the wake.

If now the cylinder begins to rotate in the negative sense with angular velocity Ω, the *major* effect of such rotation on the flow is as follows. The boundary layer on the upper surface just before separation becomes a vortex sheet with strength $V - \Omega a$, which is equal to the change in tangential velocity between its value V just outside the boundary layer and the velocity Ωa of the moving surface in the *same* direction. By contrast, the corresponding strength of the vortex sheet on the lower surface is $V + \Omega a$ because the cylinder surface is there moving in the *opposite* direction to the external flow velocity. It follows that more of the z component (eqn (369)) ζ of vorticity is shed into the wake with *positive* sign from the lower surface than is shed with negative sign from the upper surface. In short, the wake receives a significant net amount of positive vorticity, and Stokes's theorem (eqn (138)) tells us that the circulation around a closed curve C_w embracing the whole wake once in the positive sense (Fig. 69) must in consequence become positive.

On the other hand, we can apply Kelvin's theorem (eqn (147)) to a very large closed curve C which moves with the fluid and embraces the body as well as the entire wake (Fig. 69). The proof of Kelvin's theorem tells us that the circulation around C remains unchanged provided that viscous effects are negligible *along C itself*. Therefore, since the viscous forces act only in the boundary layer and wake, the circulation around a curve C outside both of these must remain unchanged (its value being zero if the motion is started from rest).

Now Fig. 69 makes clear that the circulation around C can be written as the

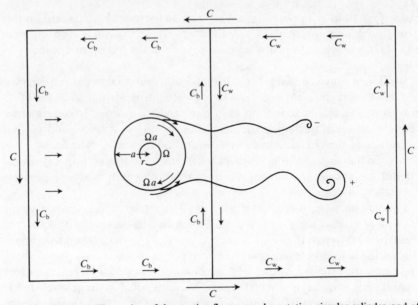

Fig. 69. Schematic illustration of the starting flow around a rotating circular cylinder, and of the associated *vortex shedding* into the wake. The circulation around a circuit C_w embracing the wake but not the body must have a positive value K. Therefore, the circulation around a circuit C_b embracing the body but not the wake has the negative value $(-K)$, as necessitated by the fact that circulation around the very large closed curve C must be zero.

sum of the circulation around C_b (which embraces the body but not the wake) and the circulation around C_w. Therefore, when a net shedding of positive vorticity into the wake gives a circulation K in the positive sense around C_w, an equal circulation K in the negative sense around the body (that is, around C_b) must simultaneously be created. This circulation, in turn, generates the flow of Fig. 68, with its associated lift, there related to the fact that external flow velocities V are increased on the upper surface of the cylinder and decreased on the lower surface. Further changes in circulation would be expected (according to these arguments, referred to earlier as representing the *major* effect of cylinder rotation) to cease as soon as such changes in external flow velocity had restored equality in the rate of shedding of vorticity from the two surfaces (such restoration would result partly from the associated changes in vortex-sheet strengths, and partly from differences in the speed of convection of shed vorticity away from the upper and lower surfaces).

The above arguments have highlighted some important principles which will be further illustrated in Section 10.2:

(i) Any process which causes a positive net rate of shedding of vorticity from a body must generate a circulation in the negative sense around the body.

(ii)   This change happens fast; that is, in the time required for shed vorticity to move well clear of the body (which at high Reynolds numbers (eqn (168)) is far less than times needed for *diffusion* over a comparable distance).

(iii)   The process is self-regulating in that the generation of circulation changes the whole external flow in ways that can eliminate the net rate of shedding of vorticity; after this the circulation can remain unchanged.

These principles, *applicable only to two-dimensional flows*, describe the major effect of cylinder rotation in changing the whole flow around the cylinder, and in generating lift.

For completeness we must, however, add that an additional, yet relatively *minor*, effect also contributes to the generation of lift on a rotating cylinder. This minor effect is, furthermore, important as the *only* effect available for generating the quantitatively much more modest lift on a rotating sphere, familiar (for example) when a tennis ball is 'sliced' from underneath.

This minor effect is that rotation acts to delay separation on the upper surface, that is to move the point of separation backwards. In Fig. 33, of course, with the solid surface at rest, the very first appearance of vorticity at the surface with the opposite sense of rotation to that in the rest of the boundary layer must cause reversed flow, as at E. However, if the solid surface is moving forwards, a larger amount of vorticity of opposite sign needs to be generated in the boundary layer before the associated negative gradients in fluid velocity generate any reversed flow. Thus, separation occurs further back, after the external flow has undergone a significantly stronger retardation.

For a tennis ball, this means that the pressure coefficients on the upper surface fall *below* those described by the broken line in Fig. 49, so as to come closer to those described by the dotted line (representing the effect of separation delayed by the different mechanism of turbulence). Evidently, some lift is caused by the pressures being less on the upper surface than on the lower.

For a rotating cylinder, by contrast, lift forces under comparable conditions are an order of magnitude greater. This is because principle (i) makes a major modification to the entire flow field. Furthermore, this modification is *enhanced* when rotation delays separation on the upper surface because this effect must reduce the external flow velocity at separation and so contribute additionally to the reduction in the rate of shedding of vorticity from the upper surface.

We may note finally that, for exceptionally high values of the circulation K around a circular cylinder, the theoretical irrotational flow takes a different form from that of Fig. 68. Thus, if

$$K > 4\pi a U, \tag{434}$$

eqn (427) has no solution and there are no stagnation points on the sphere. The calculated streamlines (on which the function ψ in eqn (428) is constant) are then as illustrated in Fig. 70, with the only stagnation point (where the streamlines, by eqn (430), must still intersect at right angles) being in the region

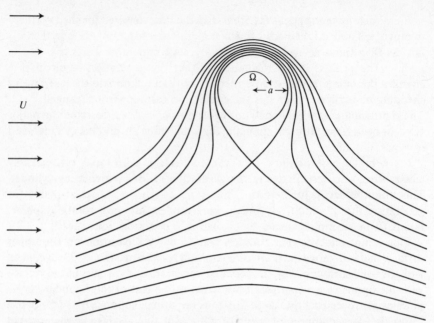

Fig. 70. Streamlines in the irrotational motion (eqn (422)) around a rapidly rotating circular cylinder of radius a when the circulation K around the cylinder takes the very large value $5\pi aU$.

of fluid below the sphere. The associated lift, which is still given by eqn (433), is enormous. Experiments have demonstrated the existence of flows similar to this, with huge lifts but also with substantial drags, in cases when the rotation speed takes values around $3U/a$ or more.

10.2 Aerofoils at incidence

In this section we analyse the flow around an aerofoil held at rest in an oncoming stream with velocity U directed at an angle of incidence α to the aerofoil's 'chord' (line joining its leading and trailing edges). This is a flow problem for which, besides exploring the mechanism that causes circulation to develop (rather as in Section 10.1), we can go farther and give a good quantitative prediction of the value at which the circulation will stabilize itself. The initial approach is that of Section 9.4, in which the mapping of eqn (395) is used to relate flows around circular cross-sections to flows around flat plates, ellipses, and aerofoils.

For a circular cross-section, evidently, the fact that the oncoming stream has been turned through an angle α cannot change the pattern of irrotational flow around the cross-section but, rather, must simply turn that whole pattern through the same angle α. This is effected if in the complex potential $f(z)$ the complex variable z is replaced by

$$ze^{-i\alpha}, \tag{435}$$

which represents a point's position relative to axes turned in the positive sense through an angle α.

Thus, from the flow without circulation given by eqn (404), we obtain in this way the flow (also without circulation)

$$f(z) = U(ze^{-i\alpha} + a^2 z^{-1} e^{i\alpha}) \tag{436}$$

which arises when the oncoming stream is at angle α to the x axis. As a check, we may note that the behaviour of

$$w = u - iv = f'(z) = U(e^{-i\alpha} - a^2 z^{-2} e^{i\alpha}) \tag{437}$$

as $|z| \to \infty$ yields the correct limiting values $u \to U \cos \alpha$ and $v \to U \sin \alpha$.

The power of the method of conformal mapping is excellently illustrated by the fact that, after the eqn (395) has mapped the trivial flow given by eqn (403) past a flat plate into the nontrivial but simple flow of eqn (404) around a circular cross-section, that flow can be rotated through an angle α as in eqn (436) and then mapped back (using eqn (395) again) in order to yield the complicated flow around a flat plate at incidence displayed in Fig. 71. Here, each streamline plotted is a locus of points (X, Y) given by eqns (397), where r and θ are linked in such a way that the imaginary part of eqn (436),

$$\psi = U(r - a^2 r^{-1}) \sin(\theta - \alpha), \tag{438}$$

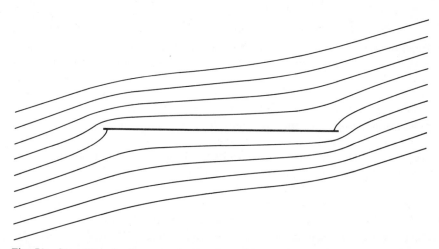

Fig. 71. Streamlines for flow around a flat plate, with zero circulation, at angle of incidence $\alpha = 20°$.

remains constant along the streamline. The case $\psi = 0$ represents the dividing streamline (on which either $r = a$ or $\theta = \alpha - \pi$ or $\theta = \alpha$), and we note that it approaches the surface of the flat plate at one stagnation point and leaves it at another, being directed normally to the surface at each as required by eqn (430). One of these stagnation points (where $\theta = \alpha - \pi$) is on the *lower* surface (since $-\pi < \theta < 0$), and the other (where $\theta = \alpha$) is on the *upper* surface (since $0 < \theta < \pi$), of the flat plate (eqns (402)).

The complex velocity in E (the domain outside the flat plate) takes the form given in eqn (388) which, with the help of eqn (383), can be written

$$w = df/dZ = (df/dz)(dZ/dz)^{-1}. \tag{439}$$

Substituting from eqns (437) and (401) we obtain

$$w = U(e^{-i\alpha} - a^2 z^{-2} e^{i\alpha})(1 - a^2 z^{-2})^{-1}. \tag{440}$$

Thus, on the flat plate (eqns (402)) which corresponds to the circle $r = a$, we have

$$z = ae^{i\theta}, \text{ giving } w = [U \sin(\theta - \alpha)](\sin \theta)^{-1}. \tag{441}$$

Equation (441) not only confirms the existence of the stagnation points (where $w = 0$) at $\theta = \alpha - \pi$ and at $\theta = \alpha$ in this irrotational flow; it also gives the much more sensational indication that w becomes *infinite* at the sharp leading edge ($\theta = \pi$) and trailing edge ($\theta = 0$) of the plate—a result which we need to understand both mathematically and physically.

Mathematically, it is evident from eqn (439) that w *must become infinite* at any point on ∂E where

$$dZ/dz = 0 \text{ but } df/dz \neq 0. \tag{442}$$

In words, then, infinite velocity must occur at a point where the mapping function $Z(z)$ from D onto E has zero derivative, *unless that point is a stagnation point for the flow in D.* For the mapping function (eqn (395)) this means that w becomes infinite at both $Z = 2a$ and $Z = -2a$ unless either of the corresponding points $z = a$ and $z = -a$ is a stagnation point of the flow in D. Of course, there is one flow (eqn (404)) in D for which both those points are stagnation points and the corresponding flow (eqn (403)) in E, far from exhibiting infinite velocity, is completely uniform. However, the rotation (expression (435)) through angle α shifts those stagnation points in D and then the corresponding value of w in E, given by eqn (441), is infinite at $\theta = 0$ and at $\theta = \pi$.

Physically, it is the fact that the leading and trailing edges of the plate have zero radius of curvature (in other words, they are *sharp* edges) which must make steady irrotational flow around those edges impossible unless the physically unrealistic condition of infinite fluid speed q, associated by eqn (431) with negatively infinite excess pressure p_e, were satisfied. Specifically, the value in expression (87) of the pressure gradient normal to any streamline requires a negatively infinite excess pressure to suck the fluid round streamlines whose radius of curvature R tends to zero as the edge is approached.

We can check this interpretation by carrying out a similar calculation for a shape with both edges given a non-zero radius of curvature, namely the ellipse given by eqns (399), obtained by applying the mapping eqn (395) to the domain D_1 outside the circle $r = c$. The flow in D_1 without circulation, given by eqn (409) for $\alpha = 0$, is converted by the substitution (435) into a flow

$$f(z) = U(ze^{-i\alpha} + c^2 z^{-1} e^{i\alpha}) \tag{443}$$

which is identical to eqn (409) except that the undisturbed stream is at a non-zero angle α to the x axis. In E_1, then, the complex velocity given in eqn (439) is

$$w = df/dZ = U(e^{-i\alpha} - c^2 z^{-2} e^{i\alpha})(1 - a^2 z^{-2})^{-1} \tag{444}$$

and, on the ellipse given by eqns (399) which corresponds to the circle given by eqns (398), this becomes

$$w = u - iv = U(e^{-i\alpha} - e^{-2i\theta} e^{i\alpha})(1 - a^2 c^{-2} e^{-2i\theta})^{-1}. \tag{445}$$

It now follows from eqn (412) that the fluid speed q on the ellipse with semi-axes A and B takes the form

$$q = |w| = U(A + B)|\sin(\theta - \alpha)|(A^2 \sin^2 \theta + B^2 \cos^2 \theta)^{-\frac{1}{2}}. \tag{446}$$

Just as for the flat plate, the points $\theta = \alpha - \pi$ and $\theta = \alpha$ are stagnation points. On the other hand, the leading edge $\theta = \pi$ and trailing edge $\theta = 0$ are now points where the body's radius of curvature takes the value $B^2 A^{-1}$, and where the fluid speed, far from becoming infinite,[†] takes the finite value

$$q = U(AB^{-1} + 1)\sin \alpha. \tag{447}$$

It is most important to understand the quite different level of significance of eqn (447) for the physical realizability of the flows around the leading edge $\theta = \pi$ and around the trailing edge $\theta = 0$ for small but non-zero values of B/A. At the leading edge, provided that $\sin \alpha$ (the sine of the angle of incidence) is limited to values not much above B/A, the fluid speed given by eqn (447) is not much greater than the undisturbed flow speed U. In consequence, the flow around the front part of the ellipse (see Fig. 72) suffers no strong retardation such as could cause boundary-layer separation. This makes a huge contrast with the case of the sharp leading edge (behind which the irrotational flow would suffer an infinitely strong retardation, thus causing the boundary layer to separate immediately). In short, *separation at a leading edge* with non-zero radius of curvature *is avoided for sufficiently small angles of incidence*.

[†] Mathematically, indeed, we know that the criterion (442) for infinite velocity cannot be satisfied in E_1 or on ∂E_1 since these do not include the points $Z = \pm 2a$ where $dZ/dz = 0$.

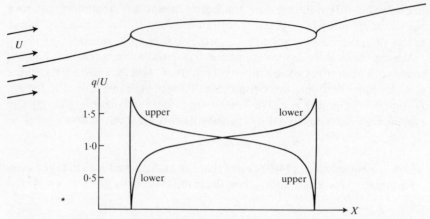

Fig. 72. The dividing streamline for flow around an elliptic cylinder of axis-ratio 0.15, with zero circulation, at angle of incidence 10°. The lower diagram plots the distribution of fluid speed q on the *upper* surface and on the *lower* surface.

There is, however, no similar effect at the trailing edge, where the distribution of fluid speed given by eqn (446) involves a strong retardation from the value in eqn (447) at $\theta = 0$ to zero at the closely neighbouring stagnation point $\theta = \alpha$. This strong retardation must cause the boundary layer to separate and there is no advantage, therefore, over the case of a sharp trailing edge (where, as before, the predicted infinitely strong retardation would cause the boundary layer to separate immediately). In this last case, indeed, the contrast between a moderate rate of shedding of negative vorticity on the upper surface, associated with the boundary layer separating a little ahead of the stagnation point at a modest value of the external flow velocity V, and a very high rate of shedding of positive vorticity close to the edge where the external flow velocity is very high, might be expected to produce major changes in the circulation K around the body which could be advantageous in certain circumstances. These arguments, tending to reinforce those of Section 9.4 in support of aerofoil shapes with rounded leading edges and sharp trailing edges, will be pursued further after the characteristics of flows at non-zero incidence α *and non-zero circulation* K (in the negative sense) have first been explored.

For a circular cross-section, of course, the appropriate complex potential is obtained from eqn (422) by applying the substitution in expression (435) so that the oncoming stream is turned through an angle α. The complex potential in D becomes

$$f(z) = U(ze^{-i\alpha} + a^2 z^{-1} e^{i\alpha}) + (iK/2\pi)\log(ze^{-i\alpha}), \tag{448}$$

and the entire streamline pattern of Fig. 68 is simply turned in the positive sense through an angle α. The stagnation points given by eqns (426) are changed, therefore, to

$$\theta = -\beta + \alpha \text{ and } \theta = \pi + \beta + \alpha. \tag{449}$$

This result (eqns (449)) has extremely interesting implications for corresponding flows obtained by the mapping eqn (395). It means that there is one particular value of β such that the condition (expression (442)) for the velocity to become infinite at the trailing edge (corresponding to the point $z = a$ or $\theta = 0$ on ∂D) is no longer satisfied, because $\theta = 0$ is itself a stagnation point in D. This occurs, by eqn (427), when

$$K = 4\pi a \, U \sin \alpha, \text{ giving } \beta = \alpha. \tag{450}$$

Thus, for one particular value (expression (450)) of the circulation K, a finite velocity should be achieved at the trailing edge in the flow obtained from eqn (448) by the mapping eqn (395).

Indeed, in the flow around a flat plate obtained in this way, the complex velocity in eqn (439) is obtained from eqn (448) as

$$w = \mathrm{d}f/\mathrm{d}Z = [U(e^{-i\alpha} - a^2 z^{-2} e^{i\alpha}) + (iK/2\pi)z^{-1}](1 - a^2 z^{-2})^{-1}. \tag{451}$$

Thus, on the flat plate (eqns (402)), corresponding to the circle $r = a$, we have

$$z = ae^{i\theta}, \text{ giving } w = U[\sin(\theta - \alpha) + \sin \beta](\sin \theta)^{-1}, \tag{452}$$

with β defined as in eqn (427). This complex velocity becomes infinite at $\theta = 0$, except when the circulation takes the particular value in expression (450). This gives $\beta = \alpha$, in which one case w can be written in a form

$$w = [U \cos(\tfrac{1}{2}\theta - \alpha)](\cos \tfrac{1}{2}\theta)^{-1} \tag{453}$$

which makes clear how w takes a very moderate finite value $U \cos \alpha$ at $\theta = 0$, even though the infinite value at the leading edge $\theta = \pi$ is still maintained. Just one stagnation point, at $\theta = \pi + 2\alpha$, is now found and the streamlines take the form shown in Fig. 73 (obtained by turning the streamlines of Fig. 68 through the angle $\beta = \alpha$ and then mapping them onto E).

These are important conclusions because the only remaining flow singularity (point where the predicted velocity is infinite) is the one at the leading edge which we know can be eliminated, by the introduction of non-zero radius of curvature, in such a way that leading-edge separation can be avoided for sufficiently small angles of incidence. At the trailing edge, on the other hand, it suggests how the self-regulation of circulation by vortex shedding (described as principle (iii) on p. 181) may, if the trailing edge remains sharp, lead to a predictable value (expression (450)) for K which can eliminate separation in that region as well.

Thus, for all values of K except that which makes $\beta = \alpha$, eqn (452) implies infinitely strong retardation between the point $\theta = 0$ of infinite velocity and

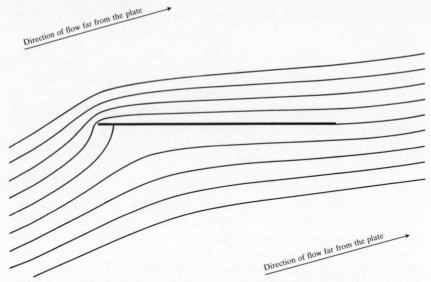

Fig. 73. Streamlines in the irrotational flow around a flat plate of length $4a$ at angle of incidence $\alpha = 15°$, when the circulation K around the plate takes the value in expression (450), which yields a finite velocity at the trailing edge.

the stagnation point at $\theta = -\beta + \alpha$. This must lead to preferential shedding of positive vorticity if $\beta < \alpha$ (so that the stagnation point is on the upper surface, as in Fig. 73) but to preferential shedding of *negative* vorticity if $\beta > \alpha$ (so that the stagnation point has moved round to the lower surface). In the first case, the circulation K in the negative sense must increase (principle (i) on p. 180), while in the second case it must decrease. Therefore, the process causes K to tend to the value in expression (450) for which $\beta = \alpha$, and for which all strong retardation tending to promote trailing-edge separation is absent.

The above arguments suggest that an *aerofoil* shape, with rounded leading edge and sharp trailing edge, avoids separation not only in symmetrical flow around the aerofoil (as Fig. 67 indicates); it may also, for sufficiently small angles of incidence, avoid separation *both* at the leading and trailing edges after the circulation has become adjusted to a value for which the velocity at the trailing edge is finite. Postponing until Section 10.3 any discussion of the beneficial consequences of this conclusion for *forces* on aerofoils, we conclude Section 10.2 with a more detailed analysis of one particular aerofoil shape (the symmetrical Zhukovski aerofoil) at incidence.

The surface of this Zhukovski aerofoil is the mapping of the circle given in eqns (417) by the mapping function (eqn (395)). The radius (eqn (414)) and centre (eqn (415)) of the circle are chosen so that it goes through just *one* point where $Z'(z) = 0$, namely the point $z = a$. The flow around the circle (eqn (417)) at angle α to the x axis and with circulation K in the negative sense has complex potential

$$f(z) = U[(z+\delta)e^{-i\alpha} + b^2(z+\delta)^{-1}e^{i\alpha}] + (iK/2\pi)\log[(z+\delta)e^{-i\alpha}] \quad (454)$$

which is obtained from eqn (448) by changing the centre and radius. The value taken by the complex velocity $w = f'(z)$ on the circle (eqn (417)) is readily calculated as

$$w = f'(z) = 2Uie^{-i\alpha}[\sin(\sigma - \alpha) + (K/4\pi bU)]. \quad (455)$$

At the trailing edge of the aerofoil, corresponding to $z = a$ or $\sigma = 0$, we know by expression (442) that the velocity must be infinite unless the corresponding point $\sigma = 0$ is a stagnation point for the flow around the circle. By eqn (455) this condition is

$$K = 4\pi bU \sin \alpha \quad (456)$$

(this is the same as in expression (450) but with the new radius b replacing a), and when this condition is satisfied the velocity from eqn (455) becomes

$$w = f'(z) = 4Uie^{-i\sigma}\sin\tfrac{1}{2}\sigma\cos(\tfrac{1}{2}\sigma - \alpha), \quad (457)$$

with stagnation points at $\sigma = 0$ and $\sigma = \pi + 2\alpha$.

By contrast, if K is zero or if it takes a positive value less than in eqn (456), eqn (455) implies a stagnation point with $0 < \sigma \leqslant \alpha$ which is mapped onto a stagnation point on the upper surface of the aerofoil, while the condition given by expression (442) implies an infinite velocity at the trailing edge itself. Between these two points, an infinitely strong retardation must generate immediate separation leading to a high net rate of shedding of positive vorticity (Fig. 74), since the boundary layer approaching the stagnation point from the lower surface must separate at a higher value of the external flow velocity V than does the boundary layer (subject to much gentler retardation) approaching it from the upper surface. Stokes's theorem (eqn (138)) then tells us that the circulation around a closed curve C_w embracing the whole wake in the positive sense must become positive.

At the same time, Kelvin's theorem (eqn (147)) tells us that the circulation around a very large circuit C embracing both body and wake must continue to take the value zero. This circulation is the sum of the circulations around C_b (embracing the body but not the wake) and C_w. It follows that a circulation K in the negative sense must start to develop around C_b and, furthermore, that this process must continue until the satisfaction of condition (456) causes the mechanism for vortex shedding to disappear.[†]

On the surface of the Zhukovski aerofoil at incidence, then, the complex

---

[†] Thus, by contrast with the mechanism for generating circulation which was illustrated in Fig. 69, that described above leads to a well defined value of K.

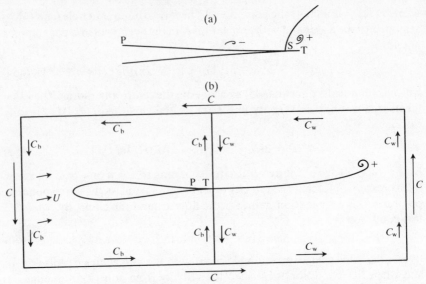

Fig. 74. Schematic illustration of the mechanism of generation of circulation in the flow around an aerofoil at positive angles of incidence.

(a)   Before circulation has been generated, separation of boundary layers in the region PT near the trailing edge (see diagram (b) for the location of this region, shown magnified in (a), with respect to the aerofoil as a whole) sheds much more positive vorticity (between T and S) than negative vorticity (between P and S).

(b)   The net positive vorticity thus shed into the wake gives a positive circulation K around the circuit C_w which embraces the wake but not the body. Therefore the circulation around a circuit C_b which embraces the body but not the wake takes the value $(-K)$. This is necessitated by the fact that circulation around the very large closed curve C must be zero. The above process continues until K takes the value given in eqn (456), leading to smooth flow at the trailing edge.

velocity $w = f'(Z)$ given by eqn (439) is obtained from the corresponding value of $w = f'(z)$ for flow around the circle which, with the condition of eqn (456) satisfied, is given by eqn (457); this only varies from the case of zero incidence in that a factor $\cos(\tfrac{1}{2}\sigma - \alpha)$ replaces $\cos(\tfrac{1}{2}\sigma)$. It follows that the fluid speed q on the surface is obtained by making the same replacement in eqn (420), giving

$$q = [U\cos(\tfrac{1}{2}\sigma - \alpha)](b^2 - 2b\delta\cos\sigma + \delta^2)b^{-1}[(b^2 - 2b\delta)\cos^2(\tfrac{1}{2}\sigma) + \delta^2]^{-1/2}. \tag{458}$$

Figure 75 plots this expression for the fluid speed on both the upper and the lower surface of the aerofoil against the X coordinate given by eqns (418) for a range of values of α. The broken lines represent values on the lower surface (except for the case $\alpha = 0$ where the same values, plotted earlier in Fig. 67, appear on both surfaces). An important inference from Fig. 75 is that for sufficiently small α (up to about 8°) these flow velocities outside the boundary

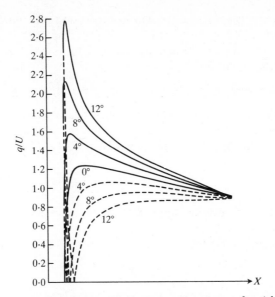

Fig. 75. The distribution (eqn (458)) of fluid speed on the upper surface (plain line) and the lower surface (broken line) of the Zhukovski aerofoil depicted in Fig. 67, for flow past the aerofoil at angles of incidence α equal to 12°, 8°, 4° and 0° (in the last case, of course, the distributions on both surfaces are identical).

layer are everywhere (even on the upper surface) only weakly retarded, so that boundary-layer separation should be avoided as predicted earlier. Then flows close to the irrotational flow with the prescribed value of circulation should be achieved. By contrast, for larger values of α (from about 10° upwards), the upper-surface flow is strongly retarded near the leading edge, suggesting in these cases a likelihood of breakdown of the irrotational flow through leading-edge separation or *stall* which is further explored in Section 10.3.

10.3 Forces on aerofoils

Those irrotational flows with circulation which *can* be maintained steadily are of special importance because they exert *lift*, that is they exert a force L at right angles to the direction of the undisturbed flow. Furthermore, the value of L, as Zhukovski discovered, has a simple relationship to the circulation K. This is the relationship in eqn (433) which was obtained in Section 10.1 for a circular cross-section, and which will now be proved valid for all shapes of section (among which, of course, aerofoil sections are the most important).

The force $(F, G, 0)$ per unit span with which any two-dimensional, steady irrotational flow acts upon a body surface of cross-section C may be written as an integral around C, in the positive sense, with respect to the arc length s measured along C. Here, an element ds of arc is related to the changes in x and y along it by the equations

$$dx = \cos \chi \, ds, \quad dy = \sin \chi \, ds, \quad \text{giving } dz = e^{i\chi} \, ds, \tag{459}$$

where χ represents the direction of that element relative to the x axis (Fig. 76). The pressure force $p_e \, ds$ per unit span acts along the inward normal (at right angles to χ) so that the force components F and G take the form

$$F = \int_C (-\sin \chi) p_e \, ds, \quad G = \int_C (\cos \chi) p_e \, ds, \tag{460}$$

where the latter form includes eqn (432) as a special case since, for the circle, $\chi = \theta + \tfrac{1}{2}\pi$ and $ds = a d\theta$.

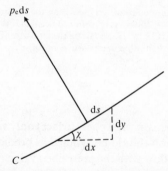

Fig. 76.   In two-dimensional flow, the force $p_e ds$ per unit span acting on an element of arc ds of a body's cross-section C acts along the inward normal in a direction specified by the unit vector $(-\sin \chi, \cos \chi, 0)$, where χ is the angle defined by eqns (459).

In these integrals the excess pressure p_e can be replaced by its steady-flow form (eqn (431)), and indeed by the simpler form $(-\tfrac{1}{2}\rho q^2)$ since the uniform pressure $\tfrac{1}{2}\rho U^2$ can generate no force. Here, q is the magnitude of the fluid velocity, whose direction is χ so that its components u, v and the complex velocity $w = u - iv$ are given by

$$u = q \cos \chi, \quad v = q \sin \chi, \quad w = qe^{-i\chi}. \tag{461}$$

Now, since w is an analytic function of z, it is desirable to combine the force integrals (eqns (460)) into the form of an integral of w^2 with respect to z which may be evaluated by Cauchy's theorem for integrals of analytic functions. In fact, using eqns (461) and (459) to substitute $-\tfrac{1}{2}\rho w^2 e^{2i\chi}$ for p_e and $e^{-i\chi} dz$ for ds in eqns (460), we obtain

$$F = \tfrac{1}{2}\rho \int_C (\sin \chi) w^2 e^{i\chi} \, dz, \quad G = -\tfrac{1}{2}\rho \int_C (\cos \chi) w^2 e^{i\chi} \, dz, \qquad (462)$$

from which we see that

$$G + iF = -\tfrac{1}{2}\rho \int_C w^2 \, dz. \qquad (463)$$

Cauchy's theorem states that the integral (in eqn (463)) of the analytic function w^2 around the closed curve C must have an unchanged value if the integral is taken around any other closed curve into which C can be deformed without passing over any singularities of the function. For externally unbounded irrotational flow, we can so deform C into a very large circle C_R of centre the origin and radius R and write[†]

$$G + iF = -\tfrac{1}{2}\rho \int_{C_R} w^2 \, dz, \qquad (464)$$

an integral which can be evaluated in terms of the circulation around C_R.

Indeed, the analytic function w in an externally unbounded domain possesses, by Laurent's theorem, the expansion

$$w = w_0 + w_1 z^{-1} + w_2 z^{-2} + \ldots \qquad (465)$$

for large $|z|$. The corresponding expansion of w^2 is

$$w^2 = w_0^2 + 2w_0 w_1 z_1^{-1} + (w_1^2 + 2w_0 w_2) z^{-2} + \ldots \qquad (466)$$

and the integral of w^2 around C_R is necessarily equal to $(2\pi i)$ times the coefficient of z^{-1} in eqn (466). Therefore, eqn (464) becomes

$$G + iF = -2\pi i \rho w_0 w_1. \qquad (467)$$

Similarly, the integral of w around C_R is $(2\pi i)$ times the coefficient of z^{-1} in eqn (465), so that

$$\int_{C_R} w \, dz = 2\pi i w_1. \qquad (468)$$

Now, the integral in eqn (468) takes the form

$$\int_{C_R} w \, dz = \int_{C_R} f'(z) \, dz = [f(z)]_{C_R} = -K, \qquad (469)$$

being the change in the complex potential $\phi + i\psi$ as we go round the circle C_R once in the positive sense, when ϕ changes by $(-K)$ (since K is the circulation

---

[†] Thus, since w^2 is analytic throughout the domain bounded by C and C_R, Cauchy's theorem states that the integral $(\int_C - \int_{C_R}) w^2 \, dz$ around that domain's boundary is zero.

in the *negative* sense), and there can be no change in ψ, by eqn (365), if there is no mass outflow across C_R. Finally, then, eqns (467) and (468) give

$$G + iF = \rho w_0 K. \tag{470}$$

Equation (470) makes it evident that, in any flow problem like that of Section 10.1 where the complex velocity (eqn (423)) takes the limiting value $w_0 = U$ for large $|z|$ (because the undisturbed flow is parallel to the x axis), the only force acting on the body is a lift

$$L = \rho U K \tag{471}$$

in the y direction (thus, $G = L$ and $F = 0$). Similarly, in flow problems like those of Section 10.2 where $w = u - iv$ takes the limiting value

$$w_0 = Ue^{-i\alpha} \tag{472}$$

for large $|z|$ because the undisturbed flow is at an angle α to the x axis, eqn (470) gives

$$G = L\cos\alpha, \quad F = -L\sin\alpha \tag{473}$$

with L as in eqn (471). This implies that the force L is directed *strictly at right angles* to the undisturbed flow. Once again, then, L is a pure lift without any drag.

This conclusion for steady irrotational flow with circulation is extremely important because of the properties of aerofoils at incidence which have been explained in Section 10.2, namely that a flow very close (outside a thin boundary layer and wake) to the irrotational flow with a definite value of the circulation can be achieved at sufficiently small angles of incidence. Such a flow should generate a substantial lift with very little drag.

For example, the Zhukovski aerofoil of Fig. 74, with circulation related to angle of incidence as in eqn (456), should achieve a lift per unit span close to

$$L = 4\pi b \rho U^2 \sin\alpha \tag{474}$$

for values of α below that upper limit when leading-edge separation or *stall* can be expected to occur. Usually, it is convenient to express lift force by means of a coefficient C_L which relates it to $\frac{1}{2}\rho U^2 S$ (as was done for drag in the eqn (256) defining C_D). On the other hand the appropriate area S is generally recognized, for a lifting wing, to be the available lifting area, that is the area of the *planform* (projection of the wing onto a horizontal plane). This means that for a two-dimensional aerofoil the corresponding area S, taken per unit span, is the *chord* h (distance between the leading and trailing edges). Thus, we define

$$C_L = L/(\tfrac{1}{2}\rho U^2 h). \tag{475}$$

For example, the chord h takes the value

$$h = 4b^2(b + \delta)^{-1} \tag{476}$$

for the Zhukovski aerofoil of eqn (418), so that the predicted value from eqn (474) for the lift implies a lift coefficient from eqn (475) equal to

$$C_L = 2\pi(1 + \delta b^{-1})\sin\alpha. \tag{477}$$

Figure 77 shows that, for those sufficiently small angles of incidence with which we are mainly concerned, the predicted behaviour (eqn (477)) of the lift coefficient (plain line) is almost a simple proportionality to α. Observed values (broken line) exhibit the same property with a slope which (because of some non-negligible effects of the thickness of the boundary layer) is slightly smaller; on the other hand, this proportionality ceases abruptly at the value of α for which leading-edge separation sets in. There is then a loss of lifting effectiveness, combined with a massive increase in drag, and the aerofoil is said

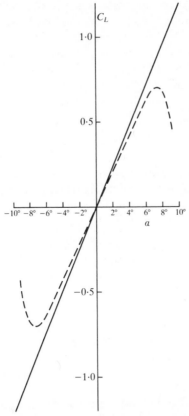

Fig. 77. The lift coefficient C_L plotted against angle of incidence α for the symmetrical Zhukovski aerofoil depicted in Fig. 67. The plain line represents irrotational-flow theory, and the broken line represents typical measured values.

to have stalled. The behaviour of stalled aerofoils is complex and interesting but is outside the scope of the present book with its prime emphasis on assisting in the understanding of flows that are efficient for engineering purposes.

Any symmetrical aerofoil like that of Figs. 75 and 77 must evidently have properties which are symmetrical between positive and negative angles of incidence. Thus, it is equally capable of generating positive and negative lifts, as Fig. 77 shows. We shall see (in Chapter 11) that such a capability is important for certain purposes; thus, an aircraft's fin needs an aerofoil section of this type. The wing of an aircraft, on the other hand, needs to generate mainly *positive* lifts, to balance the aircraft's weight.

These considerations led, at an early period in aviation, to the use of aerofoil sections with the property known as 'camber' which has the effect of raising the whole graph of C_L versus α above the level shown in Fig. 77; thus, increasing the positive values of C_L that are achievable without stalling, at the expense of giving up the ability to achieve significant negative values. Without describing modern methods in aerofoil design, we here show how a very minor modification of the process used to generate the symmetrical Zhukovski aerofoil (by mapping the points in eqn (417) of a certain circle) produces a *cambered Zhukovski aerofoil* with these properties.

For this purpose it is, indeed, only necessary to map the points

$$z = -\delta + i\varepsilon + be^{i\sigma}, \tag{478}$$

which form a circle with centre $z = -\delta + i\varepsilon$ *above* the centre $z = -\delta$ of eqn (417). To obtain a sharp trailing edge the circle produced by eqn (478) must pass through the point $z = a$ where the mapping function eqn (395) satisfies $Z'(z) = 0$, so that

$$b^2 = (a+\delta)^2 + \varepsilon^2. \tag{479}$$

Figure 78(a) shows this circle (for a particular choice of δ/b and ε/b), together with the shapes of the dividing streamline at three angles of incidence, covering the range of angles for which irrotational flow about the corresponding aerofoil is closely achievable. In each case, the circulation must adjust itself so that the stagnation point at which the dividing streamline leaves the surface is at $z = a$; then by expression (442) a finite velocity is achieved at the trailing edge of the corresponding aerofoil (Fig. 78(b)).

The important feature of cambered aerofoils is that the flow at zero incidence (with the undisturbed stream parallel to the chord) is *already* a flow with substantial circulation K and therefore substantial lift. In this flow (see the broken line in Fig. 78) the leading edge of the aerofoil is itself a stagnation point, so that there is no tendency whatsoever for rapid flow around the leading edge to pose any threat to separation. This threat begins, rather, at a certain positive angle of incidence (dash-dotted line) for which the stagnation

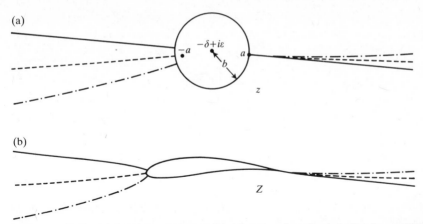

Fig. 78. Use of the conformal mapping (eqn (395)) to generate from the circle given by eqn (478) (with $\delta/b = 0.1$ and $\varepsilon/b = \sin 6°$) a cambered aerofoil, and to derive the shapes of the dividing streamlines in flow around it at angles of incidence $\alpha = -6°$ (plain line), $\alpha = 0°$ (broken line), and $\alpha = 6°$ (dash-dotted line).

(a)   Dividing streamlines are shown for irrotational flow past the circle with the circulation chosen in each case to make $z = a$ (where $\sigma = -6°$) a stagnation point, so that mapping on to the Z-plane gives a finite velocity at the aerofoil's trailing edge. For $\alpha = -6°$, a flow with zero circulation and with a straight dividing streamline satisfies this condition (plain line). For $\alpha = 0°$, we need the circulation K to take the value $4\pi bU \sin 6°$ (broken line). For $\alpha = 6°$, we need $K = 4\pi bU \sin 12°$ (dash-dotted line).

(b)   Corresponding shapes of dividing streamlines in the Z plane, where the circle is mapped onto a cambered Zhukovski aerofoil.

point has moved substantially away from the leading edge on the aerofoil's lower surface. However, the lift at this angle of incidence is almost double that achievable with the corresponding symmetrical aerofoil, through addition of the lift associated with zero incidence. On the other hand, at a certain negative incidence (plain line), where the flow is without circulation and therefore without lift, a threat of separation is posed by the flow from a stagnation point on the *upper* surface of the aerofoil.

Figure 78 explains, in short, how cambered aerofoils may increase the range of positive lifts attainable, at the expense of loss of the capability to achieve negative lifts. Figure 79, indicating how the curve of C_L against α is changed from that of Fig. 77, confirms that cambered aerofoils are able to exhibit this frequently desirable combination of properties.

This section, like Section 8.4, is concluded by a calculation of the *moment* exerted on a body in steady irrotational flow. Nevertheless, the aim of such a calculation is somewhat different for two-dimensional flows where the fluid pressures exert a resultant force L on the body. Instead of determining a

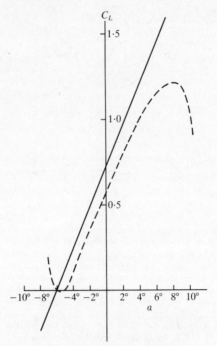

Fig. 79.   The lift coefficient C_L plotted against angle of incidence α for the cambered Zhukovski aerofoil depicted in Fig. 78. The plain line represents irrotational-flow theory, and the broken line represents typical measured values.

destabilizing couple as in Section 8.4, we are concerned rather to establish the *line of action* of the lift L.

In the integrals (eqns (460)) for the x and y components of force, we can obtain the moment about the origin for each element of force per unit span p_e dS through multiplying its y component by x and subtracting the result of multiplying its x component by y. The total moment per unit span is therefore

$$M = \int_C (x \cos \chi + y \sin \chi) \, p_e \, dS. \tag{480}$$

Here, the term in brackets may be written

$$\mathrm{Re}\left[(x + iy)(\cos \chi - i \sin \chi) \right] = \mathrm{Re}\,(z e^{-i\chi}), \tag{481}$$

where Re stands for 'real part of'. Thus, M is the real part of an integral identical with that obtained for $G + iF$ except that it contains an additional factor z; this changes eqn (463) into

$$M = -\tfrac{1}{2}\rho \, \mathrm{Re} \int_C z w^2 \, dz. \tag{482}$$

When Cauchy's theorem is applied to obtain the integral in eqn (482) by evaluating it around a large circle on which eqn (466) holds, we obtain

$$M = -\tfrac{1}{2}\rho \operatorname{Re}\left[2\pi i\,(w_1^2 + 2w_0\,w_2)\right], \tag{483}$$

where the term in round brackets represents the coefficient of z^{-1} in zw^2. Actually, eqns (468) and (469) show w_1^2 to be purely real so it makes no contribution to eqn (483), which therefore becomes

$$M = -2\pi\rho \operatorname{Re}(iw_0 w_2). \tag{484}$$

Equation (484) makes it easy to calculate the moment of the force per unit span acting on the aerofoils studied earlier in this chapter. The calculation is assisted by the fact that when the function (eqn (395)) is used to map D onto E the form of the expansion (eqn (465)) for large z in D or for large Z in E is the same up to and including the term in w_2.

For the symmetrical Zhukovski aerofoil at incidence, then, we can expand the value of $w = df/dZ$ given, by eqns (439) and (454), as

$$w = \left\{U\left[e^{-i\alpha} - b^2(z+\delta)^{-2}\,e^{i\alpha}\right] + (iK/2\pi)(z+\delta)^{-1}\right\}(1 - a^2 z^{-2})^{-1} \tag{485}$$

to give

$$w_0 = Ue^{-i\alpha},\; w_1 = (iK/2\pi),\; w_2 = U(a^2 e^{-i\alpha} - b^2 e^{i\alpha}) - (iK\delta/2\pi), \tag{486}$$

so that the moment in eqn (484) (with K substituted from eqn (456)) becomes

$$M = -2\pi\rho U^2\,(a^2 + b\delta)\sin 2\alpha. \tag{487}$$

This *negative* moment about the origin implies that the line of action of the lift force intersects the aerofoil's chord in a point which lies upstream of the aerofoil's mid-point. To a close approximation this is the aerofoil's 'quarter-chord point', in the sense that its distance from the leading edge is about one quarter of the chord h (that is, of the distance between the leading and trailing edges).

By contrast, the lift force associated with the cambered Zhukovski aerofoil at *zero* incidence is found to act almost through the mid-chord point of the aerofoil (as, indeed, the flow pattern indicated by the plain line in Fig. 78 would suggest). Finally, at positive angles of incidence, the lift of such a cambered aerofoil is found to be in two parts: a part associated with the camber that acts almost through the mid-chord point; and a part associated with incidence that acts almost through the quarter-chord point.

11
Wing theory

This final chapter is devoted to the three-dimensional flow around *wings*, in the sense of long surfaces whose cross-sections at right angles to their length are aerofoil sections, but where those sections' dimensions and shape may vary gradually along the length of the wing. Our first and principal objective is to analyse conditions in which wings are *aerodynamically efficient*, in the sense that they can generate large lifts at the expense of very much smaller drags. Then (in the concluding Section 11.4) we briefly illustrate several areas of advantageous application of such aerodynamic efficiency.

Within three-dimensional flows around wings, the mechanism (Section 10.2) which generates circulation around an aerofoil section is still found to operate effectively. However, rather than producing just an *irrotational* flow with circulation, it is found to generate certain residual vorticity, called *trailing vorticity*. The theory of *vortex dynamics* (Chapter 5) is, in Section 11.1, pursued somewhat further in order to clarify the properties of trailing vorticity.

Of course, only irrotational motions can be generated from rest by impulses applied at their boundaries (Chapters 6 and 8), yet it nevertheless proves fruitful to consider the total impulse that would theoretically be needed to generate a motion with vorticity, even though some of that impulse would have to be applied in the midst of the fluid. In fact, the rather simple expression for the impulse of a vortex system as so defined (Section 11.2) proves invaluable for the study of forces on wings. This study, while elucidating the effectiveness of wings for the production of large lift and small drag, demonstrates, however, that the vortex system necessarily generates a certain additional element of drag (additional, that is, to the drag associated with the non-zero thickness of the boundary layer and the associated defect of volume flow in the wake). It is, indeed, this additional element, known as the *induced drag* (Section 11.3), whose minimization represents the last mathematical problem addressed in this book.

11.1 Trailing vorticity

In the steady three-dimensional flow around a wing at a moderate angle of incidence, the local motion around each aerofoil section must qualitatively be subject to the same powerful mechanism as was illustrated in Fig. 74 for the generation of circulation around the section. Quantitatively, of course, the level at which the circulation takes up a steady value may be influenced by overall features of the three-dimensional flow; for example, we could perhaps foresee (correctly, as will appear in Section 11.3) that this overall flow might modify the *effective* angle of incidence of the local motion around the aerofoil section. Furthermore, we must be ready to consider the possible nature and consequences of any *variation* in the level of circulation around different aerofoil sections, particularly since their dimensions and shape may vary gradually along the length of the wing.

In this context, we note too that at the very tip of the wing the circulation must have fallen to zero. In fact, a closed curve just beyond the wing tip (Fig. 80) must consist of particles of fluid that have reached this position

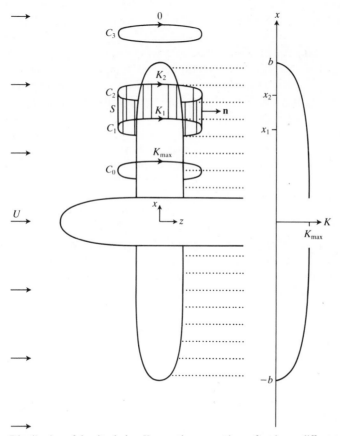

Fig. 80.   Distribution of the circulation K around cross-sections of a wing at different positions x along its span. The largest circulation K_{max} is that taken around a curve C_0 embracing the wing close to the fuselage. Zero circulation, however, appears around a curve C_3 beyond the wing tip. The circulation around curves (such as C_1 and C_2) at intermediate positions takes values gradually decreasing from K_{max} to zero. The dotted lines indicate trailing vorticity, with its strength determined by applying Stokes's theorem to a collar-shaped region S spanning C_1 and C_2.

without being subject to any of those viscous forces whose action is limited to the boundary layer and wake, so that Kelvin's theorem (eqn (147)) applies to such a closed curve and the circulation around it must be zero. It follows that any gradual variation in the level of circulation around different aerofoil sections that is achieved in steady flow must take that circulation K from its maximum value K_{max} all the way to zero at the wing tip (Fig. 80). We now investigate the implications of such variation as far as the ultimate steady state is concerned, while returning at the end of the section to the mechanisms that may be responsible for generating this steady state.

The coordinates used in this chapter's studies of three-dimensional flow revert to those of Chapter 7 (from eqn (224) onwards), in that the z direction is taken as the direction of the undisturbed stream. (See the discussion following eqn (367) for an explanation of why different coordinate systems are appropriate to two- and three-dimensional flow.) Thus, the z direction is the direction of any drag force acting on the wing, while the y direction is taken as that of the lift force and the x direction points 'to starboard' along the length of the wing (Fig. 80) (note that the theory of wings will here be given detailed development only for wing pairs symmetrical about the plane $x = 0$ and extending from the port wing tip $x = -b$ to the starboard wing tip $x = b$). These coordinates form a right-handed system, and the expected circulation K around the wing is in the *positive* sense about the x axis.[†] Changes of incidence are considered as being made by tilting the wing in the given undisturbed stream, rather than by keeping the body surface fixed and varying the stream direction as in Chapter 10.

In the above coordinate system we consider the circulations, say K_1 and K_2 with $K_2 < K_1$, around closed curves C_1 and C_2 which embrace two of the wing's different aerofoil sections: those at $x = x_1$ and at $x = x_2 > x_1$ (Fig. 80). Then, taking the collar-shaped surface S (also embracing the wing) as spanning the closed curves C_1 and C_2, we apply Stokes's theorem (eqn (134)) with the normal $\mathbf{n}$ to S pointing outwards from the wing. This gives

$$\int_S \boldsymbol{\omega} \cdot \mathbf{n}\, dS = \int_{\partial S} \mathbf{u} \cdot d\mathbf{x} = K_1 - K_2, \tag{488}$$

because the boundary ∂S when taken in the positive sense relative to $\mathbf{n}$ consists of the two closed curves C_1 and C_2, followed respectively in the positive and negative senses about the x direction (Fig. 80).

Now, we expect a 'streamlined' wing shape to achieve irrotational flow everywhere except in a thin boundary layer and wake. In that case non-zero vorticity $\boldsymbol{\omega}$ can appear on a surface S that is well clear of the boundary (Fig. 80)

[†] The reason why the same circulation K in the coordinate system of Chapter 10, with x replacing z, was in the *negative* sense (understood to mean the negative sense about the z axis in that system) is that replacing z by x while keeping the system right-handed requires the z direction to point 'to port' rather than to starboard along the length of the wing.

only in that thin strip where it intersects the wake. We take this intersection to be at right angles so that $z =$ constant on the strip, where also $x_1 < x < x_2$, and where (similarly) y is confined to the thickness of the wake (say, $y_1 < y < y_2$). Then $\boldsymbol{\omega} \cdot \mathbf{n}$ in eqn (488) is the z component of vorticity ζ and the equation becomes

$$\int_{x_1}^{x_2} k(x)\,dx = K_1 - K_2, \tag{489}$$

where

$$k(x) = \int_{y_1}^{y_2} \zeta\,dy \tag{490}$$

is the integral of the vorticity component ζ across the thickness of the wake.

Equation (489) shows that any *difference in the circulations* about adjacent aerofoil sections requires the existence of *trailing vorticity* with a non-zero integral (eqn (490)) across the wake. In the nomenclature of Section 5.4 this makes the wake a *vortex sheet*, with its *strength* $k(x)$ defined as the integral (eqn (490)) of the vorticity across the thickness of the sheet. More strictly, the strength of a vortex sheet is a vector, since the vorticity $\boldsymbol{\omega}$ is, but we shall see that the z component (in eqn (490)) of that vector is necessarily its dominant component.

In the limit as $x_2 - x_1$ becomes small, eqn (489) takes the form

$$k(x) = -K'(x), \tag{491}$$

where $K(x)$ is the circulation around the aerofoil section at position x. Thus, as $K(x)$ decreases continuously from its maximum value $K_{\max}$ to zero at the wing tip (Fig. 80), the slope $-K'(x)$ of that decrease is necessarily equal to the strength $k(x)$ of the trailing vortex sheet.

This conclusion (another major discovery of Prandtl) reconciles the apparent discrepancy between (i) the procedures of Chapter 10, allowing a certain choice of the circulation K in the two-dimensional irrotational flow around an aerofoil section, and (ii) the demonstration in Chapter 8 that, for *any* finite body in three dimensions (including, of course, a wing), the irrotational flow around it is unique. We see that the mechanisms of Section 10.2 *are* free to act to determine the circulations $K(x)$ over the different aerofoil sections of a three-dimensional wing, but that $K(x)$ must decrease to zero at the wing tips and the overall motion, rather than being irrotational, *must incorporate a trailing vortex sheet* whose strength is given by the slope (eqn (491)) of that decrease.

In order to demonstrate that the strength of the trailing vortex sheet is dominated by the component ζ of vorticity in the z direction (that is, in the direction of the undisturbed stream) we need to pursue the subject of *vortex dynamics* a little further than was attempted in Chapter 5. For this purpose we return once more to comparing the components in expressions (66) and (67) of the two vectors $\mathbf{u} \cdot \nabla \mathbf{u}$ and $\nabla(\frac{1}{2}\mathbf{u} \cdot \mathbf{u})$. Subtracting expression (67) from expression (66), we obtain

$$w\eta - v\zeta, \quad u\zeta - w\xi \tag{492}$$

for the first two components of the difference (with a similar expression $v\xi - u\eta$ for the third component), where ξ, η, ζ are the components in expression (105) of the vorticity $\boldsymbol{\omega}$. We conclude that

$$\mathbf{u} \cdot \nabla \mathbf{u} - \nabla(\tfrac{1}{2}q^2) = \boldsymbol{\omega} \times \mathbf{u}, \tag{493}$$

a result compatible with two previous conclusions: eqn (176) which puts the left-hand side of eqn (493) equal to zero when $\boldsymbol{\omega} = 0$ (that is, for irrotational flow); and eqn (81) which under all circumstances gives that left-hand side zero component in the direction of $\mathbf{u}$.

Equation (493) allows us to rewrite the equation of motion (eqn (175)) in the form

$$\partial \mathbf{u}/\partial t + \boldsymbol{\omega} \times \mathbf{u} = -\nabla(\tfrac{1}{2}q^2 + \rho^{-1} p_e); \tag{494}$$

which, it is worth remarking, has an interesting interpretation from the standpoint of general mechanics. Figure 20, describing the motion of a fluid particle P as a combination of a rigid rotation at angular velocity $\frac{1}{2}\boldsymbol{\omega}$ with irrotational motions (translation plus pure rate of strain), makes it clear that in a system of *rotating axes*, rotating about C with angular velocity $\frac{1}{2}\boldsymbol{\omega}$, the motion of P is irrotational. Now, the left-hand side of eqn (494) represents the well known form for the rate of change of the vector $\mathbf{u}$ at C in such axes rotating with angular velocity $\frac{1}{2}\boldsymbol{\omega}$ (the term $\boldsymbol{\omega} \times \mathbf{u}$ being the *Coriolis* acceleration), where this rate of change, in such axes for which the motion is locally irrotational, must according to eqn (177) take the form shown on the right-hand side.

In the steady flows with which this section is concerned, eqn (494) may be written

$$\boldsymbol{\omega} \times \mathbf{u} = -\operatorname{grad}(\tfrac{1}{2}q^2 + \rho^{-1} p_e), \tag{495}$$

which we now apply to the thin-vortex-sheet wake. Outside it, and indeed throughout the irrotational part of the flow, eqn (179) assures us that $\frac{1}{2}q^2 + \rho^{-1} p_e$ has a constant value. Thus, even though this quantity is always expected to take reduced values within the middle of any wake, its *net change across the wake must be zero*. Thus, the gradient on the right-hand side of eqn (495) has a component across the wake which averages to zero. But the vector product of two vectors in the plane of the vortex-sheet wake would, if it were non-zero, be directed across the wake. Therefore the average value of $\boldsymbol{\omega} \times \mathbf{u}$ in the wake must be zero.

The above arguments confirm that, at least in an average sense, $\boldsymbol{\omega}$ must be well aligned with $\mathbf{u}$ in the wake: *trailing vorticity is streamwise vorticity*. Thus, the z component in eqn (490) of vortex-sheet strength is necessarily its leading component. This inference remains correct, furthermore, even though the direction of **u** varies a little from that of the undisturbed stream (the z direction). For example, the x component of **u** certainly changes across the vortex sheet (Section 5.4) by an amount equal to its strength $k(x)$, being directed towards the wing tip below the sheet and away from the tip above it (Fig. 80); but this change is necessarily small compared with the main z component of fluid flow.

In the flow around a wing with circulation, the overall pattern of *additional vorticity* (additional, that is, to the vorticity which would be present in the boundary layer and wake of a flow without any circulation around sections of the wing) includes not only trailing vorticity but also *attached* or *bound vorticity*. Figure 81 shows how Stokes's theorem (eqn (134)), applied in the plane $x = $ constant (with the normal **n** in the x direction) to a region S between the boundary of the aerofoil section (where the no-slip condition **u** $= 0$ applies) and a closed curve C outside the boundary layer, gives

$$\int_S \xi \, dS = \int_C \mathbf{u} \cdot d\mathbf{x} = K(x). \tag{496}$$

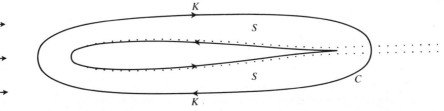

Fig. 81. Stokes's theorem, applied to a region S between the boundary of an aerofoil and a closed curve C which embraces the aerofoil outside the boundary layer, shows that the integral of the x component ξ of vorticity over the entire boundary layer is equal to the circulation K around C. Thus, the total *additional vorticity* (additional, that is, to the pattern of vorticity when the circulation is zero) is equal to K.

In fact, the vorticity's x component ξ takes positive values in the boundary layer on the wing's upper surface, and negative values on the lower surface. Equation (496) shows that all of these integrate to *zero* when the circulation $K(x)$ is zero, but that the pattern of additional vorticity has a positive integral $K(x)$ over the region S when $K(x)$ is positive. About half of that additional vorticity is present in the boundary layer on the *upper* surface (which, as a vortex sheet, has increased strength when $K(x) > 0$), and about half on the lower surface (whose vortex-sheet strength, of opposite sign, has reduced magnitude when $K(x) > 0$). In short, the additional vorticity includes a total amount $K(x)$ of bound vorticity (attached to both boundary layers) in addition to trailing vorticity in a vortex sheet of strength $-K'(x)$.

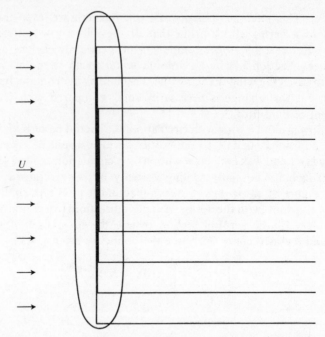

Fig. 82. Schematic illustration of the overall pattern of additional vorticity for flow around a lifting wing.

Figure 82 gives a schematic (though rather crude) sketch of this overall pattern of additional vorticity, taking a broad macroscopic view of the bound vorticity by modelling it as a concentrated line vortex of gradually varying strength $K(x)$ which is represented in the diagram by the thickness of the line. As the line's thickness is reduced by losing strands, these strands turn through 90° to become individual line-vortex elements of a trailing vortex sheet of strength $-K'(x)$. The pattern shown emphasizes the solenoidality of the vorticity field, and has been picturesquely described as a combination of *horseshoe vortices*.

Postponing until Section 11.3 any attempt to give a more refined picture of the steady-flow pattern of additional vorticity, we conclude this section with an indication of how the trailing vorticity is initially generated after a flow has been started from rest. Figure 74 showed how circulation is generated around each aerofoil section through the shedding of some vorticity, the circulation around which is equal and opposite. This vorticity is depicted in the figure after a lapse of time during which the general flow has carried it clear of the aerofoil. For the flow around the wing as a whole, Fig. 83 gives a sketch (schematic, as in Fig. 82) of the flow when a similar time has elapsed since it was started from

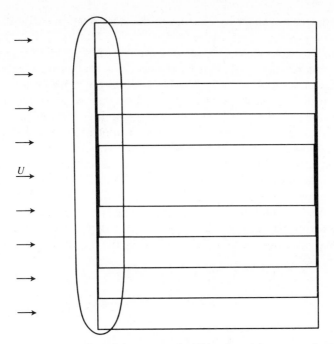

Fig. 83. Schematic illustration of the pattern of additional vorticity at an early stage in the development of flow around a lifting wing.

rest. Behind a section where the circulation assumes the value $K(x)$, the circulation around the shed vorticity must be $-K(x)$. Taking once again a broad macroscopic view, this time of the shed vorticity by modelling it as a concentrated line vortex, we can once more represent its gradually varying strength $-K(x)$ by changes in the thickness of the line. The resulting 'starting pattern' (Fig. 83) re-emphasizes the solenoidality of the vorticity field, a property which, indeed, necessitates the generation of trailing vorticity so as to link *shed vorticity of variable strength* to the bound vorticity (of equal and opposite variable strength) that is still attached to the wing.

11.2 Impulse of a vortex system

A good understanding of the forces on wings can be derived from a study of the *impulse* required to set up any vortex system; including, of course, the vortex systems described in Section 11.1. This study is found to demand a major extension of the ideas used in Section 8.3 (for irrotational flow); nevertheless, a link between impulse and far-field dipole strength is again identified. We precede the general study by illustrating the close relations between certain special vortex systems and dipole fields.

Consider, for example, the flow associated with a line vortex of strength K in the form of a *rectangle* ABCD (Fig. 84) (note also that such rectangular line vortices are basic elements of the flow pattern of Fig. 83). We know that, when any line vortex is present, a single-valued velocity potential for the flow must always possess a surface of discontinuity; thus, a line vortex in the form of a *straight* line can be described by the velocity potential (eqn (348)) with a surface of discontinuity $\theta = \pi$. Such a discontinuity is necessary because along any closed curve embracing the line vortex the velocity potential (eqn (202)) must increase by exactly the circulation K. These considerations suggest that, for the line vortex of Fig. 84, it should be possible to find a velocity potential ϕ which is single-valued but possesses a *rectangular* surface of discontinuity ABCD across which the potential changes by exactly K (matching the amount by which ϕ increases on rounding any one of the rectangle's four sides).

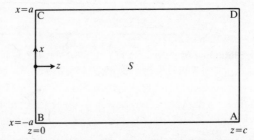

Fig. 84. The line vortex of strength K in the form of a rectangle ABCD can be analysed by use of a velocity potential ϕ, with a rectangular surface of discontinuity ABCD across which ϕ changes by K. This ϕ is also the potential of a uniform distribution of dipoles of strength as given by expression (499) per unit area over ABCD.

Such a potential may easily be constructed, by considering the surface of discontinuity as the limit of a thin rectangular layer of very small thickness $2h$, lying between $y = -h$ and $y = +h$. Then a very large flow velocity $(0, V, 0)$ in the thin layer is needed in order to bring about a substantial change

$$2hV \tag{497}$$

in the potential ϕ between $y = -h$ and $y = +h$. This flow velocity can be readily generated by a uniform distribution of point sources over the side $y = -h$ of the rectangular layer, which yields a total volume flow V per unit area, and which is exactly matched on the side $y = +h$ by a uniform distribution of point sinks to absorb that volume flow. In the limiting case

$$h \to 0, \quad V \to \infty, \quad 2hV \to K \tag{498}$$

the velocity potential acquires the necessary discontinuity K across the rectangular surface ABCD.

Evidently, this limiting process leads to a uniform distribution of *dipoles* over the rectangular area. The dipole strength per unit area is

$$(0, -K, 0) \tag{499}$$

being given by eqn (292) with $\mathbf{h} = (0, -h, 0)$ and $J = V$.

At first sight it may seem surprising that the rectangular line vortex has the *same* velocity potential as a distribution of dipoles over the rectangular area it encloses, with the dipole strength directed normal to that area and having a magnitude per unit area equal to the strength of the line vortex. However, since the two potentials have identical discontinuities across the rectangular surface, their difference is a velocity potential which is continuous everywhere, and the argument in eqn (295) can be used to prove that its gradient is identically zero. We may, furthermore, reinforce this conclusion by checking as follows that at a point (x, y, z) *very close to one side of the rectangle*, say $x = a$, $y = 0$, the potential of the dipole distribution does approximate to the potential in eqn (348) of a line vortex in the form of that straight line. In fact,

$$\int_{-\infty}^{a} (Ky/2\pi)[(x-X)^2 + y^2]^{-1} \, dX = (K/2\pi) \arctan[y/(X-a)]; \tag{500}$$

where the integrand is the potential of a line dipole on $x = X$, $y = 0$ with strength $K dX$ per unit length, and the right-hand side is the form of eqn (348) appropriate to a line vortex along $x = a$, $y = 0$. Note that, *sufficiently close to that vortex*, the integral in eqn (500) is not significantly modified when we replace the real lower limit of integration (say, $-a$) by $-\infty$ or when, similarly, the integrand used is, as in eqn (328), the potential of a line dipole extending between $z = -\infty$ and $z = +\infty$ (rather than between, say, $z = 0$ and $z = c$ as it does in reality).

In the far field, of course, the entire distribution of dipoles of strength (as in expression (499)) per unit area is 'seen' as a single dipole of strength

$$(0, -SK, 0), \tag{501}$$

where S is the area of the rectangle. This expression (area enclosed times vortex strength) for the far-field dipole strength of a rectangular line vortex is of value both in its own right and as a guide to what the 'impulse' of the system may turn out to be.

Often, it is useful to express expression (501) in terms of the *moment*

$$\int \mathbf{x} \times \boldsymbol{\omega} \, dV \tag{502}$$

of the vorticity distribution over all space, where, however, only the *interior* of a line vortex contributes to this integral. In fact, the part of a line vortex stretching from x to $x + dx$ contributes an amount

$$x \times (K\,dx) \tag{503}$$

to expression (502), where ω is in the direction dx and where eqn (152) gives us its magnitude integrated over the cross-sectional area of the line vortex as K. For a rectangular line vortex in the plane $y = 0$, the vector product (expression (503)) can be written

$$(0, K(z\,dx - x\,dz), 0), \tag{504}$$

and its integral (expression (502)) comes to *twice* the dipole strength (expression (501)) since the area S enclosed by the rectangle can be written in both the forms

$$S = \int x\,dz = -\int z\,dx. \tag{505}$$

For the line vortex of Fig. 84, then, the far-field dipole strength can be written

$$\mathbf{G} = \tfrac{1}{2} \int_L \mathbf{x} \times \boldsymbol{\omega}\,dV \tag{506}$$

over any region L including all the vorticity. In the rest of this section we demonstrate the correctness of the result (eqn (506)) for very general vorticity distributions, while also relating **G**, rather as in Section 8.3, to the *impulse* of the system.

In this general discussion we allow for the presence of a solid boundary S (which might, for example, be the surface of a wing). We study the externally unbounded flow outside S, but assume that *vorticity* is entirely confined to a finite region L bounded internally by S and externally by a surface Σ (Fig. 85).

In the spirit of Section 6.3, furthermore (see eqn (197) in particular), we are concerned only with the velocity field $\mathbf{u}_0$ which remains after that irrotational flow which satisfies the boundary conditions has been subtracted out.[†] Similarly, we are concerned only with the vorticity ω which remains after subtraction of the vorticity in that thin vortex sheet (the boundary layer) which reconciles the said irrotational flow with the no-slip condition on S; we are concerned, in short, with *additional* vorticity (Section 11.1). On these definitions we have

$$\text{curl } \mathbf{u}_0 = \boldsymbol{\omega} \tag{507}$$

and can take $\mathbf{u}_0$ as satisfying the no-slip condition

$$\mathbf{u}_0 = 0 \text{ on } S \tag{508}$$

in addition to the condition that the magnitude of $\mathbf{u}_0$ tends to zero far from the body.

[†] Thus, any impulse calculated for the velocity field $\mathbf{u}_0$ constitutes an *addition* to the impulse associated with the irrotational flow.

Equation (171) proves that a motion started from rest by impulses applied at its boundary must be an *irrotational* motion with the velocity potential from eqn (173). This conclusion, however, does not rule out the theoretical possibility that the rotational velocity field $\mathbf{u}_0$ might be generated instantaneously from rest by a distribution of impulses,

$$\rho \mathbf{u}_i \text{ per unit volume,} \tag{509}$$

applied directly to particles of fluid within L. Under these circumstances eqn (171) would be replaced by

$$\rho \mathbf{u}_0 = \rho \mathbf{u}_i - \int_0^\tau (\operatorname{grad} p_e)\,d\tau, \tag{510}$$

where the right-hand side is the sum of the directly applied impulse per unit volume and an additional impulse per unit volume associated with any very large excess pressures p_e acting during the very short time τ of application of the impulse. Equation (510) can be written

$$\mathbf{u}_0 = \mathbf{u}_i + \operatorname{grad} \Phi \text{ with } \Phi = -\rho^{-1} \int_0^\tau p_e\,dt, \tag{511}$$

where, instead of using ϕ as in eqn (173), we prefer to use Φ in the present case (where it is not actually a velocity potential).

It is, indeed, only in the region outside Σ (see Fig. 85) that the velocity field $\mathbf{u}_0$ is given to be without vorticity, so that it assumes the irrotational form

$$\mathbf{u}_0 = \operatorname{grad} \phi_0 \text{ outside } \Sigma. \tag{512}$$

It follows that expression (511) can be satisfied with

$$\mathbf{u}_i = 0 \text{ and } \Phi = \phi_0 \text{ outside } \Sigma, \tag{513}$$

and therefore that the externally applied impulse (expression (509)) can be confined entirely to the region L inside Σ.

Using an important idea introduced by Horace Lamb (1849–1934), we define Φ in the region L inside Σ as the solution of Laplace's equation satisfying the boundary conditions

$$\Phi = \phi_0 \text{ on } \Sigma \text{ and } \Phi = 0 \text{ on } S, \tag{514}$$

where the method based on eqns (186) to (190) readily shows Φ to be uniquely determined in L by these boundary conditions (expressions (514)). Along with expression (513), these conditions constrain Φ to be continuous across Σ; this implies that, while its normal derivative $\partial \Phi/\partial n$ will in general be discontinuous across Σ, *its derivatives in tangential directions must be continuous*.

Therefore, if the distribution $\rho \mathbf{u}_i$ per unit volume of impulses applied within L is defined by eqn (511), with Φ specified as in expressions (512) and (514), then the tangential component of $\mathbf{u}_i$ must be continuous across Σ with the *zero*

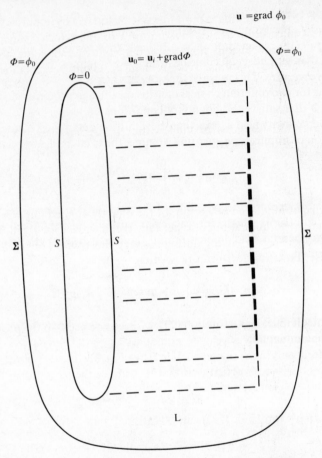

Fig. 85. Illustrating a vortex system confined to a region L bounded internally by a body surface S and externally by a surface Σ (outside which the flow is irrotational). Such a motion may be generated instantaneously from rest by a system of impulses (expression (509)), the total applied impulse being given by eqn (521) as $(\frac{1}{2}\rho)$ times the moment of the vorticity distribution.

value (expression (513)) that it assumes outside Σ. Also on S, eqns (508) and (514) imply zero tangential components for both $\mathbf{u}_0$ and grad Φ, and so also for $\mathbf{u}_i$. Summing up,

$$\mathbf{u}_i \times \mathbf{n} = 0 \text{ on } \partial L, \tag{515}$$

where the boundary ∂L consists of the surfaces S and Σ. We may note also that

$$\text{curl } \mathbf{u}_i = \text{curl } \mathbf{u}_0 = \boldsymbol{\omega}, \text{ div } \mathbf{u}_i = \text{div } \mathbf{u}_0 = 0, \tag{516}$$

since grad Φ is given as having zero divergence and curl.

By using eqn (511) (with a well defined choice of Φ) to specify $\mathbf{u}_i$ we are undoubtedly specifying a distribution of impulses (expression (509)) which *can* generate the given motion $\mathbf{u}_0$. Indeed, they certainly will generate it provided that the chosen Φ coincides with $(-\rho^{-1})$ times the time-integral of the excess pressures. This must be the case, furthermore, since for given $\mathbf{u}_i$ it is impossible for two different choices of Φ in eqn (511) to produce velocity fields $\mathbf{u}_i + \operatorname{grad} \Phi$ that satisfy the boundary conditions on $\mathbf{u}_0$.

The given velocity field $\mathbf{u}_0$ can, then, be set up instantaneously from rest by a distribution of impulses $\rho \mathbf{u}_i$ per unit volume, the total applied impulse being

$$\mathbf{I} = \rho \int_L \mathbf{u}_i \, dV. \tag{517}$$

We may now explore the possibility of a relationship between $\mathbf{I}$ and the quantity (eqn (506)) which, for particular vorticity distributions, has been identified as their far-field dipole strength and which, by the expressions (516), can be written

$$\mathbf{G} = \tfrac{1}{2} \int_L \mathbf{x} \times \operatorname{curl} \mathbf{u}_i \, dV. \tag{518}$$

The result (eqn (305)) for irrotational flow suggests that it may be profitable to calculate the difference

$$\rho \mathbf{G} - \mathbf{I}. \tag{519}$$

Its x component, for example, can be written

$$\tfrac{1}{2}\rho \int_L \left[y\left(\frac{\partial v_i}{\partial x} - \frac{\partial u_i}{\partial y}\right) - z\left(\frac{\partial u_i}{\partial z} - \frac{\partial w_i}{\partial x}\right) - 2u_i \right] dV$$

$$= \tfrac{1}{2}\rho \int_L \left[\frac{\partial(yv_i)}{\partial x} - \frac{\partial(yu_i)}{\partial y} - \frac{\partial(zu_i)}{\partial z} + \frac{\partial(zw_i)}{\partial x} \right] dV$$

$$= \tfrac{1}{2}\rho \int_{\partial L} \left[y(v_i n_x - u_i n_y) - z(u_i n_z - w_i n_x) \right] dS, \tag{520}$$

which is zero since the vector product (eqn (515)) of $\mathbf{u}_i = (u_i, v_i, w_i)$ and of $\mathbf{n} = (n_x, n_y, n_z)$ is zero on ∂L. Similarly the other components of expression (519) are zero, which allows us to draw the extremely useful conclusion that the *total impulse* required to set up the velocity field $\mathbf{u}_0$ from rest, with vorticity ω, is

$$\mathbf{I} = \rho \mathbf{G} = \tfrac{1}{2}\rho \int_L (\mathbf{x} \times \mathbf{\omega}) \, dV, \tag{521}$$

which is $\tfrac{1}{2}\rho$ *times the moment of the vorticity distribution.*

We may verify, too, that in this general case $\mathbf{G}$ is the far-field dipole strength (expression (296)) of the irrotational flow field existing outside Σ. This is because eqns (521), (517), and (511) give

$$\mathbf{G} = \rho^{-1}\mathbf{I} = \int_L \mathbf{u}_i \, dV = -\int_L (\text{grad } \Phi) \, dV + \int_L \mathbf{u}_0 \, dV = \mathbf{G}_1 + \mathbf{G}_2. \quad (522)$$

Here, in relation to the irrotational flow outside Σ, the second integral $\mathbf{G}_2$ in eqn (522) clearly takes the same form $V\mathbf{U}$ as in eqn (301); where the region D inside Σ (comprising both the stationary body and the region L of moving fluid) has volume V while its centroid is moving with velocity $\mathbf{U}$. Again, the first integral $\mathbf{G}_1$ can be written

$$-\int_{\partial L} \Phi \mathbf{n} \, dS = -\left(\int_\Sigma + \int_S \right) \Phi \mathbf{n} \, dS = -\int_\Sigma \phi_0 \mathbf{n} \, dS \quad (523)$$

by expression (514); where $\mathbf{n}$ is the outward normal from L so that eqn (523) coincides with the definition (eqn (296)) of $\mathbf{G}_1$ (where the normal $\mathbf{n}$ is taken *inwards* into D). Thus, eqn (522) establishes $\mathbf{G}$ as the far-field dipole strength.

These extremely general results owe much of their importance to the fact that in a changing motion the force $\mathbf{F}$ with which the body acts on the fluid is still given as the rate of change of impulse

$$\mathbf{F} = d\mathbf{I}/dt, \quad (524)$$

even though the proof given earlier for the same result (eqn (306)) in the irrotational case[†] is inapplicable when the impulse is no longer applied at the solid boundary. To establish the general truth of eqn (524) in a changing flow with vorticity, we suppose the velocity field at time t_1 to have impulse $\mathbf{I}_1$, while the velocity field at a very slightly later time t_2 has impulse $\mathbf{I}_2$. Then the motion is capable of having been started from a state of rest at time $t_1 - \tau$ by the application of the impulse $\mathbf{I}_1$ in a very short time τ. A force $\mathbf{F}(t)$ is then applied by the body on the changing fluid flow for $t_1 < t < t_2$. Finally, an impulse $-\mathbf{I}_2$ is capable of restoring the fluid to rest at time $t_2 + \tau$.

We now prove eqn (524) by considering the momentum of the fluid bounded externally by a very large sphere Σ_R of centre the origin and radius R. None of this fluid has any momentum at time $t_1 - \tau$ or at the later time $t_2 + \tau$ (when it is again brought to rest). On the other hand we can write the change of momentum during this period as

$$\mathbf{I}_1 + \int_{t_1}^{t_2} F(t) \, dt - \mathbf{I}_2 \quad (525)$$

(representing the external impulses applied between $t_1 - \tau$ and t_1, between t_1 and t_2 and between t_2 and $t_2 + \tau$) plus the effects of any pressure forces

---

[†] Actually, the result is even more important for a system of vortices, whose position is necessarily changing by Helmholtz's theorem.

acting over the spherical surface Σ_R or of momentum convected across Σ_R. However, such effects can be made as small as we please by taking R large enough. Thus, in the change of momentum

$$\int_{t_1-\tau}^{t_2+\tau} (-p_e\,\mathbf{n}\,dS)\,dt = \int_{t_1-\tau}^{t_2+\tau} (\partial\phi/\partial t + \tfrac{1}{2}q^2)(\rho\mathbf{n}\,dS)\,dt \qquad (526)$$

associated with the excess pressure p_e acting on a small element dS of Σ_R, the integral of $\partial\phi/\partial t$ is exactly zero (since the fluid is at rest at times $t_1 - \tau$ and $t_2 + \tau$) while the quantity $\tfrac{1}{2}q^2$ is of order R^{-6} for large R (as, too, is any effect of momentum convection across dS) so that its integral over Σ_R can be made arbitrarily small for large enough R. Finally, then, the expression (525) must be zero, in agreement with eqn (524).

To sum up the conclusions of this section, any vortex system possesses a far-field dipole strength $\mathbf{G}$ given as half the moment of the vorticity distribution. The impulse of the vortex system is $\rho\mathbf{G}$ and, in a changing flow, the rate of change of total impulse[†] represents the total force with which any solid boundary is acting on the fluid. For a line vortex in the form of a rectangle, $\mathbf{G}$ has the value in expression (501), directed at right angles to the plane of the rectangle and with magnitude equal to its area S times the strength K of the line vortex.

11.3 Lift and induced drag

The above considerations help to clarify further the relationship between lift and circulation which was established for two-dimensional flows in Section 10.3 by a method based on Cauchy's theorem. The same value (see eqn (471)) of the lift per unit span in two-dimensional flow is, indeed, readily derived as the rate at which the *impulse per unit span* is changing.

Thus, of the additional vorticity in Fig. 74, the shed vorticity (about which the circulation is K in the positive sense) is steadily increasing its *distance* from the bound vorticity (about which the circulation is K in the negative sense) which is attached to the surface of the aerofoil. Now, such distance, of which the rate of increase is U, represents the *area per unit span* embraced by the vortex system. Thus, the magnitude ρSK of the downward impulse $\mathbf{I} = \rho\mathbf{G}$ has a value per unit span which increases at the rate ρUK. This force with which the aerofoil acts on the fluid is directed at right angles to the undisturbed stream, and is necessarily equal and opposite to the *lift ρUK* with which the fluid acts upon unit span of the aerofoil.

The main importance of arguments along these lines, however, is that they can be applied not only to two-dimensional flows around aerofoil sections but

---

[†] A total which, in general, includes the impulse of the vortex system and also the impulse of the irrotational flow that satisfies the boundary conditions.

also to three-dimensional flows around wings. The immediate three-dimensional analogue of Fig. 74 is the crudely schematic Fig. 83, indicating the pattern of vorticity a short time after the flow around a wing has been started from rest. The corresponding analogue of the argument just used to associate circulation with lift would lead to a value

$$L = \rho U \int_{-b}^{b} K(x)\,dx \tag{527}$$

for the total lift on the wing. This result, which is consistent with the idea that the lift per unit span is $\rho U K$, would follow from the pattern of vorticity in Fig. 83 being analysed in a regrouped form as in Fig. 86. Here, between planes $x = $ constant a distance dx apart, the bound vorticity and shed vorticity with equal and opposite circulations $K(x)$ around them are regarded as elements of a line vortex of strength $K(x)$ which takes the form of a rectangle with sides dx (in the x direction) and Ut (in the z direction), where t is the time since the vorticity was shed. This rectangle's area $Ut\,dx$ increases at a *rate* $U\,dx$, so the impulse of the line vortex increases at a rate $\rho(U\,dx)\,K(x)$, thus leading to eqn (527) for the rate of change of impulse of the whole pattern of vortices.

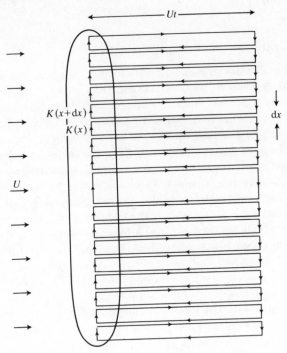

Fig. 86. Regrouping of the pattern of additional vorticity shown in Fig. 83 into an arrangement of line vortices in the form of narrow rectangles.

Evidently, the cumbersome regrouping of the vorticity pattern in Fig. 86 is equivalent to the more natural arrangement in Fig. 83. Thus, the strength of the combined line vortex present where one side of the above rectangular line vortex coincides with a side of the adjacent line vortex (also rectangular) of strength $K(x + dx)$ is

$$K(x) - K(x + dx) = -K'(x)\,dx = k(x)\,dx, \qquad (528)$$

in agreement with eqn (491) for the strength of the trailing vortex sheet.

Nevertheless, a direct calculation of rate of change of impulse for a pattern of additional vorticity such as is illustrated in Figs. 83 (for the case of a flow recently started from rest) and 82 (for a well established flow) is undoubtedly desirable. This requires a calculation of the *impulse of the streamwise trailing vorticity per unit length*, an expression which, upon multiplication by U (the rate of increase of length) must give the required rate of change of impulse.

Whether for each 'horseshoe vortex' in Fig. 82 or for each rectangular vortex in Fig. 83, its area S per unit of streamwise length is equal to its *width*; that is, to the difference between the values of x on its two trailing-vortex elements of equal and opposite strength $k(x)\,dx$. The value per unit length of the impulse (given in general as ρSK) is therefore obtained by *adding* the values of $\rho x k(x)\,dx$ for the two elements. Thus, the whole pattern of trailing vorticity has impulse

$$\rho \int_{-b}^{b} xk(x)\,dx \qquad (529)$$

per unit length. The associated lift is

$$L = \rho U \int_{-b}^{b} xk(x)\,dx, \qquad (530)$$

this being the rate of change of downward impulse in the said pattern (of which an additional length U per unit time is being generated).

Actually, the two alternative expressions (530) and (527) for the lift on a wing are both found to be useful. Furthermore, they are readily proved through integration by parts to be identical, since $k(x) = -K'(x)$ and since the circulation $K(x)$ vanishes at the wing tips $x = \pm b$ (Section 11.1).

When a wing achieves an upward lift, then, it is acting with an equal and opposite downward force on the fluid to create a trailing vortex wake with a substantial downward impulse (expression (529)) per unit length. A first approximate idea of the type of motion in this trailing vortex wake can be obtained from *the two-dimensional motion* (in the (x, y) plane) *associated with the same pattern of vorticity*, taken as a plane vortex sheet of strength $k(x)$. Figure 87 shows that two-dimensional motion for a particular vorticity distribution $k(x)$. The velocity potential ϕ_T of this two-dimensional wake model, which is obtained by combining potentials from eqn (348) for vortices

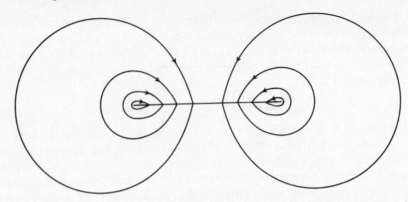

Fig. 87. Illustration, by means of a plot of its streamlines, the two-dimensional motion, with velocity potential ϕ_T, associated with a given pattern of trailing vorticity $k(x)$. The horizontal component of velocity changes discontinuously by an amount $k(x)$ across the vortex sheet.

of strength $k(X) dX$ at $x = X$, $y = 0$, so that $\tan \theta = y/(x - X)$, is

$$\phi_T = \int_{-b}^{b} [k(X)/2\pi] \arctan [y/(x - X)] dX. \tag{531}$$

The overall appearance of the motion is seen to be fully consistent with the fact that a *downward* impulse would be needed to set it up.

Beyond the first approximation given by the two-dimensional wake model of Fig. 87, it is possible to move along two quite different lines in order to obtain a more accurate picture of movements in the wake:

(i) by adopting a three-dimensional model as in Fig. 82, with trailing vortices linked through bound vortices to make a horseshoe-vortex pattern (rather than a two-dimensional model which, effectively, assumes the trailing vortices to be indefinitely extended parallel lines);

(ii) by a theory allowing the property that vortex lines move with the fluid (Helmholtz's theorem) to cause the shape of the vortex sheet to be distorted by the flow pattern of Fig. 87.

Before quantitatively proceeding along the very necessary course (i), we give a brief qualitative account of the interesting but rather less important effect mentioned under (ii).

This effect (ii) is commonly described as a 'rolling up' of the vortex wake. Figure 88 shows sections of the wake by planes $z = $ constant at increasing distances downstream. It indicates first of all how newly shed trailing vorticity in the form of a plane is capable of being distorted by a pattern of flow like that of Fig. 87, which permits the possibility of greater downward motion at the centre than at the tips. Next, it shows how the same vorticity (which by now has been swept further downstream from the wing) would be further distorted,

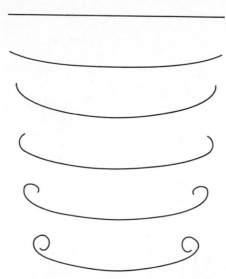

Fig. 88. 'Rolling up' of a trailing vortex wake. Different sections of the wake by vertical planes are sketched 'in elevation', with those sections' *distances behind the wing* increasing down the page.

especially at the tips, by an inward component of movement induced by the lower vortices. Further downstream still, the vortex sheet curls up still further. Finally, it rolls up into a pair of *relatively* narrow line vortices, whose equal and opposite strengths each have magnitude

$$\int_0^b k(x)\,dx = -\int_0^b K'(x)\,dx = K(0) \tag{532}$$

equal to the total strength of that half of the vortex sheet which has positive strength $k(x)$. If the distance between the centres of these rolled up vortices is $2B$, the impulse per unit length has become $2\rho BK(0)$ so that the lift is U times this, giving

$$B = L/2\rho U K(0). \tag{533}$$

By eqns (530) and (532) this can also be written as

$$B = \int_0^b xk(x)\,dx \Big/ \int_0^b k(x)\,dx, \tag{534}$$

an equation which locates the rolled up vortex of positive strength at the *centroid* of the initial distribution of positive trailing vorticity. Figure 89 shows these same developments in 'plan' form.

It is *both* because the above processes are only complete far downstream, *and* because the redistribution is not such as to affect substantially the vorticity

Fig. 89. Sketch in 'plan' form of the rolling up of a trailing vortex wake (the direction of flow being down the page).

pattern's far field (including motions induced in the neighbourhood of the wing itself),[†] that the aerodynamics of most wing shapes can be satisfactorily analysed by methods which (like those given below) ignore the rolling up of trailing vorticity. The latter effect needs, however, to be appreciated if commonly observed phenomena, including aircraft condensation trails, are to be properiy understood.

It is the equivalence of any rectangular line vortex to a uniform distribution (expression (499)) of dipoles over the enclosed area which especially facilitates the study (see (i) above) of the flow due to *a pattern of horseshoe vortices*

[†] For example, the far-field *dipole strength* is unaltered.

(Fig. 82). This equivalence implies a velocity potential

$$\phi = \int_{-b}^{b} \frac{k(X)}{4\pi} \left[\arctan\left(\frac{y}{x-X}\right) + \arctan\left\{ \frac{y[(x-X)^2+y^2+z^2]^{\frac{1}{2}}}{(x-X)z} \right\} \right] dX,$$

$$(535)$$

which may readily be calculated as the combined potential of area distributions of downward pointing dipoles, each distribution being of uniform strength $(0, -k(X)dX, 0)$ per unit area and filling the area $-X < x < X, y = 0, z > 0$; that is, an area in the form of a rectangle which has been extended *indefinitely* at its downstream end, and is bounded by a line vortex of strength $k(X)dX$ at $x = X$ and one of equal and opposite strength $k(-X)dX$ at $x = -X$. Specifically, the potential of a single horseshoe vortex is calculated as the sum of the contributions to eqn (535) from X and from $-X$, and is obtained from the potential (eqn (294)) of a dipole as the double integral over the above area

$$[yk(X)dX/(4\pi)] \int_{-X}^{X} dx_1 \int_{0}^{\infty} [(x-x_1)^2 + y^2 + (z-z_1)^2]^{-\frac{3}{2}} dz_1$$

with respect to the dipole's position $(x_1, 0, z_1)$.

Note that, when z is large and positive, the two arctan terms in eqn (535) become equal so that the potential coincides with the two-dimensional form in eqn (531). In fact, for each X, this happens when

$$z \gg [(x-X)^2 + y^2]^{\frac{1}{2}},$$

which, indeed, is the condition that at the point (x, y, z) the line vortex on $x = X, z > 0$ is seen as indefinitely extended in both directions. Such a two-dimensional asymptotic behaviour for large positive z is to be expected on the 'horseshoe vortex' model (which, of course, neglects any rolling up of the vortex sheet). For large negative z, on the other hand, and specifically when

$$z \ll -[(x-X)^2 + y^2]^{\frac{1}{2}},$$

the two arctan terms in eqn (535) cancel out; this corresponds to the absence of any vorticity upstream of the wing.

By contrast, in the immediate neighbourhood of the wing (where z is relatively small), the two arctan terms in eqn (535) represent two distinct phenomena: the effects of the trailing vorticity and the effects of the bound vorticity respectively. Note that the first arctan term, which represents the effect of trailing vorticity, makes a contribution $\frac{1}{2}\phi_T$ equal to *exactly half of the two-dimensional wake potential* (eqn (531)), as might be expected since the indefinitely extended line vortices generating ϕ_T can be thought of as combinations of line vortices stretching *both* from $z = 0$ to $z = +\infty$ *and* from $z = -\infty$ to $z = 0$ (of which only the former is present, although each would generate comparable flows for small z).

In the meantime, the contribution to eqn (535) from the second arctan term when y and z are both small is approximately

$$\left[\int_{-b}^{x}\frac{k(X)}{4\pi}\,dX - \int_{x}^{b}\frac{k(X)}{4\pi}\,dX\right]\arctan\frac{y}{z},$$

since that second arctan term is approximately $\arctan(y/z)$ for $X < x$ and $-\arctan(y/z)$ for $X > x$. The expression (536) with $k(x) = -K'(x)$ can be written

$$-[K(x)/2\pi]\arctan(y/z), \tag{536}$$

which represents the local two-dimensional motion associated with the bound vorticity on the 'horseshoe vortex' model which regards it as concentrated in a line vortex along the x axis. Really, of course, that bound vorticity is distributed through boundary layers, and its effect is accompanied by that of the circulationless irrotational flow about the aerofoil section so that the portion (expression 536)) of the potential ϕ on a simple horseshoe-vortex model must be regarded as representing the true two-dimensional flow in the (z, y) plane of that aerofoil section with circulation $K(x)$ around it. To that local two-dimensional flow of the type studied in Chapter 10 the above model adds locally another flow $\frac{1}{2}\phi_{T}$, which is also two-dimensional but in the (x, y) plane (Fig. 87).

For flow around wings, then, the local two-dimensional motion generated near where each aerofoil section disturbs the uniform stream $(0, 0, U)$ may be regarded as just a first approximation which, locally, can be raised to the level of a *second approximation* if we add on the motion $\frac{1}{2}\phi_{T}$ produced there by the trailing vorticity. The local effect of this additional motion is in two parts, which are associated respectively with its x and y components $\frac{1}{2}\partial\phi_{T}/\partial x$ and $\frac{1}{2}\partial\phi_{T}/\partial y$.

Of these, the former produces an interesting *spanwise* component of motion over the wing. Its velocity $\frac{1}{4}k(x)$ is directed towards the wing tips on the *lower* surface and away from the wing tips on the *upper* surface;[†] however, this motion produces no significant dynamical effect since it is in no way opposed by the wing's presence.

By contrast, the additional y component $\frac{1}{2}\partial\phi_{T}/\partial y$ of the local flow represents a motion in the y direction which is *continuous* (and, normally, negative) and which is significantly disturbed by the aerofoil. Thus, it combines with the oncoming uniform stream (of velocity U in the z direction) to *reduce the effective angle of incidence* of the flow around the aerofoil (Fig. 90).

In fact, it is easily shown that the value of $\partial\Phi_{T}/\partial y$ given by eqn (531) is a continuous, even function of y which for small y can be closely approximated

---

[†] Here $k(x)$, being the strength of the vortex sheet in Fig. 87, represents the change in the x component of velocity across it in the motion defined by ϕ_{T}.

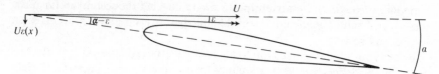

Fig. 90. The trailing vortex wake produces a downward component $U\varepsilon(x)$ of flow in the neighbourhood of an aerofoil section. This combines with the oncoming stream, of velocity U at angle of incidence α, to make a resultant (double arrow) such that the effective angle of incidence is changed to $\alpha - \varepsilon$.

by its value for $y = 0$. This value is usually negative and may be written

$$(\partial \phi_{\mathrm{T}}/\partial y)_{y=0} = -2U\varepsilon(x), \tag{537}$$

where $\varepsilon(x)$ is small. The resultant of the undisturbed stream with velocity $(0, 0, U)$ and of the local flow $(0, -U\varepsilon, 0)$ generated by trailing vorticity represents a motion tilted downwards at a small angle ε to the z axis. Then, if the geometric angle of incidence of the wing in the undisturbed stream is α, its effective angle of incidence to the local motion in the immediate environment of the aerofoil is reduced to

$$\alpha - \varepsilon(x). \tag{538}$$

This conclusion affects, to a relatively modest extent, the level of circulation $K(x)$ which the flow around the aerofoil tends to assume: it becomes that which, in two-dimensional flow, would be associated with the angle of incidence $\alpha - \varepsilon(x)$ rather than with α. This, in turn, affects the lift per unit span $\rho U K(x)$. Suppose, for example, that the lift coefficient (eqn (475)) in two-dimensional flow around the aerofoil section were observed to take the form

$$C_L = a_0[\alpha + \alpha_0(x)]. \tag{539}$$

Here measurements (see, for example, Fig. 77) usually yield a value of a_0 around 6 (as against a value slightly greater than 2π suggested by theoretical results such as eqn (477) which ignore the thicknesses of boundary layers), while for a symmetrical aerofoil $\alpha_0 = 0$ but for a cambered aerofoil α_0 is defined so that the observed plot of C_L against α (Fig. 79) intersects the axis $C_L = 0$ at $\alpha = -\alpha_0$. Then, since C_L takes the form $K/(\frac{1}{2}Uh)$ in terms of the circulation K, the relation between $K(x)$ and the chord $h(x)$ of the corresponding aerofoil section may be written

$$K(x) = \tfrac{1}{2}Uh(x)a_0[\alpha - \varepsilon(x) + \alpha_0(x)]. \tag{540}$$

Here, the term $\varepsilon(x)$ represents a modest correction to $K(x)$ and hence to the overall lift (eqn (527)). Admittedly, the value of $\varepsilon(x)$ defined by eqn (537) does itself depend on the distribution of $K(x)$, but it is straightforward to solve eqn (540) through an iterative process where $\varepsilon(x)$ is first calculated from the uncorrected values of $K(x)$, and then obtained more accurately from corrected values, and so on.

Beyond this modest correction to lift, however, the reduction $\varepsilon(x)$ in effective angle of incidence generates a quantitatively much more important effect which results from the fact that the force $\rho U K(x)$ per unit span acting on the aerofoil is directed *at right angles to the flow* (Section 10.3). When, therefore, the local motion around the aerofoil is tilted downwards through an angle $\varepsilon(x)$, the force $\rho U K(x)$ becomes tilted backwards through the same angle so that it includes not only a lift but also a small element of drag, equal to

$$\rho U K(x)\varepsilon(x) \tag{541}$$

per unit span. Because the trailing vorticity 'induces' the reduction $\varepsilon(x)$ in effective angle of incidence, this is usually called an element of *induced drag*. The overall induced drag is

$$D_i = \rho U \int_{-b}^{b} K(x)\varepsilon(x)\,dx. \tag{542}$$

Thus, whereas in any *two-dimensional* flow which avoids premature separation of boundary layers the drag is exclusively that associated with defect of volume flow in a thin wake, *three-dimensional* flows over lifting wings experience in addition the induced drag in eqn (542).

We can understand this distinction in terms of the different character of the vortex wakes in the two cases. When a two-dimensional aerofoil at incidence is caused by an external force to move through fluid at rest this force must do work to set up the 'starting vortex' motion (Fig. 74) but no further work needs to be done thereafter to maintain that vortex motion. For a three-dimensional wing, on the other hand, the trailing vorticity pattern has a certain *kinetic energy per unit length* which turns out to have exactly the form in eqn (542). Thus, the element of external thrust needed to overcome the induced drag D_i does an amount of work (eqn (542)) per unit distance which *coincides with* the work needed to produce the kinetic energy per unit length of new trailing vortex wake.

In order to verify the statement just made we write down the kinetic energy of the two-dimensional wake motion ϕ_T per unit length in the form

$$\tfrac{1}{2}\rho \int_{C} \phi_T(\partial\phi_T/\partial n)\,ds, \tag{543}$$

where the closed curve C consists of the upper and lower sides of the vortex sheet. Equation (543) is obtained (exactly like eqn (309) in three dimensions) by applying the Divergence Theorem to the vector $\phi_T \,\text{grad}\,\phi_T$ in the domain between C and a circle C_R of very large radius R. The normal $\mathbf{n}$ points outwards from this domain so that $\partial\Phi_T/\partial n$ takes the form $-\partial\phi_T/\partial y$ on the upper side, and $+\partial\phi_T/\partial y$ on the lower side, of the vortex sheet. Accordingly, the kinetic energy per unit length (eqn (543)) can be written

$$-\tfrac{1}{2}\rho \int_{-b}^{b} K(x)(\partial\phi_T/\partial y)_{y=0}\, dx, \tag{544}$$

where $K(x)$ is the difference between the values of ϕ_T on the upper and lower sides of the vortex sheet; and, by eqn (537), this is precisely the eqn (542) for D_i as stated above.

The last mathematical problem addressed in this book is that of determining the minimum possible value that the induced drag (eqn (542)) can take for a wing of given span $2b$ which develops a lift L (given by eqn (527)) in a stream of velocity U. It is equivalent to the problem of determining the least possible kinetic energy per unit length D_i for a two-dimensional wake flow ϕ_T whose impulse per unit length is L/U. As with so many other two-dimensional problems (Chapters 9 and 10) this one can be treated expeditiously by the method of conformal mapping.

In order to avoid the reintroduction of z as a complex variable in a chapter where z is needed as the third Cartesian coordinate, we rewrite the conformal mapping eqn (395) as

$$Z = z_T + (\tfrac{1}{2}b)^2 z_T^{-1}. \tag{545}$$

This, as shown in Fig. 91, maps the circle

$$r = \tfrac{1}{2}b, \text{ where } z_T = x_T + iy_T = re^{i\theta} \tag{546}$$

onto the upper and lower surfaces of the vortex sheet

$$-b \leqslant x \leqslant b, \quad y = 0, \text{ where } Z = x + iy \tag{547}$$

(compare Section 9.4). The potential ϕ_T is the real part of a complex potential f which is an analytic function of Z and which the mapping allows us to consider also as an analytic function of z_T.

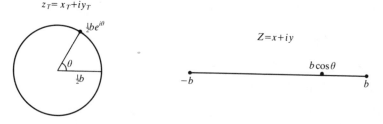

Fig. 91. The two-dimensional motion ϕ_T, associated with a pattern of trailing vorticity $k(x)$ for $-b \leqslant x \leqslant b$, $y = 0$, is conveniently studied by means of the conformal mapping (eqn (545)) of the region outside a circle of radius $\tfrac{1}{2}b$ onto the region outside the trailing vortex wake.

Here, Laurent's theorem states that an expansion

$$f = f_1 z_T^{-1} + f_2 z_T^{-2} + f_3 z_T^{-3} + \ldots \tag{548}$$

in inverse powers of z_T is possible outside the circle given by expression (546); but the symmetry of Fig. 87 about the x axis tells us that ϕ_T must be an odd function of y and hence also of y_T, so that all the coefficients f_n must be pure imaginary. Writing

$$f_n = \tfrac{1}{2}i(\tfrac{1}{2}b)^n K_n, \tag{549}$$

where K_n is real, we obtain

$$f = \tfrac{1}{2}i \sum_{n=1}^{\infty} (\tfrac{1}{2}br^{-1})^n K_n(\cos n\theta - i \sin n\theta). \tag{550}$$

Thus, on the circle given by expression (546), the real part ϕ_T of the complex potential in eqn (550) satisfies

$$\phi_T = \tfrac{1}{2} \sum_{n=1}^{\infty} K_n \sin n\theta \tag{551}$$

and

$$\partial\phi_T/\partial r = -b^{-1} \sum_{n=1}^{\infty} nK_n \sin n\theta. \tag{552}$$

Now, the circulation $K(x)$ represents the difference between the values of ϕ_T on the upper surface of the vortex sheet (corresponding to $0 < \theta < \pi$) and those on the lower surface $(-\pi < \theta < 0)$. This difference, by eqn (551), is

$$K(x) = \sum_{n=1}^{\infty} K_n \sin n\theta, \tag{553}$$

where, by eqn (545) and expression (546),

$$x = b \cos \theta. \tag{554}$$

Thus, the K_n are coefficients in a Fourier Sine Series expansion of the circulation $K(x)$ with respect to the variable θ defined by eqn (554). In particular, the lift given by eqn (527) can be written

$$L = \rho Ub \int_0^{\pi} K(x) \sin \theta \, d\theta = \tfrac{1}{2}\pi\rho UbK_1 \tag{555}$$

in terms of the standard expression for the first Fourier coefficient K_1.

Also, using eqn (439), we obtain

$$(\partial\phi_T/\partial y)_{y=0} = |dZ/dz_T|^{-1}(\partial\phi_T/\partial r)_{r=\frac{1}{2}b} = (2 \sin \theta)^{-1}(\partial\phi_T/\partial r)_{r=\frac{1}{2}b}. \tag{556}$$

Therefore, by eqn (552), the definition (eqn (537)) of $\varepsilon(x)$ gives

$$\varepsilon(x) = (4Ub \sin \theta)^{-1} \sum_{n=1}^{\infty} nK_n \sin n\theta, \tag{557}$$

which is a convenient expression for the angle $\varepsilon(x)$, by which the flow around each aerofoil section is tilted downwards, in terms of the coefficients K_n defined above. Furthermore, the induced drag (eqn (542)) can, after the substitution given by eqn (554), be written

$$D_i = \rho U \int_0^\pi K(x)\varepsilon(x)b \sin \theta d\theta \qquad (558)$$

and eqns (553) and (557) allow us to rewrite this as

$$D_i = \tfrac{1}{4}\rho \int_0^\pi \left(\sum_{n=1}^\infty K_n \sin n\theta \right)\left(\sum_{n=1}^\infty nK_n \sin n\theta \right) d\theta = \tfrac{1}{8}\pi\rho \sum_{n=1}^\infty nK_n^2 \quad (559)$$

in terms of the coefficients K_n.

Equation (559) provides the solution to our minimization problem. For given lift L, eqn (555) fixes K_1 but leaves the other coefficients free. Now, for fixed K_1, the minimum value of eqn (559) is $\tfrac{1}{8}\pi\rho K_1^2$, a minimum that is attained when all the other coefficients are zero. Thus, for given lift L, specified by eqn (555), the minimum induced drag is

$$(D_i)_{\min} = (2\pi\rho b^2)^{-1}(L/U)^2, \qquad (560)$$

a result which we are also free to regard as specifying the *minimum kinetic energy* D_i *per unit length* for a trailing vortex wake of given impulse L/U per unit length and given span $2b$.

The minimum induced drag is achieved when the distribution (eqn (553)) of circulation along the span $-b \leqslant x \leqslant b$ takes the form

$$K(x) = K_1 \sin \theta = K_1 (1 - x^2 b^{-2})^{\frac{1}{2}}; \qquad (561)$$

which is commonly described as an *elliptic* distribution. This is the way in which the circulation must fall gradually from its maximum value K_1 to zero at the tips (with the lift, $\rho U K(x)$ per unit span, varying in direct proportion to it) if the total lift is to be achieved with minimum induced drag. Taken alongside eqn (540) for $K(x)$ this consideration is found to be rather important for wing design.

Note that in eqn (540) *the reduction $\varepsilon(x)$ in effective angle of incidence takes a value independent of x,* given by eqn (557) as $(4Ub)^{-1}K_1$, under conditions of minimum induced drag.[†] Substituting this value of ε in eqn (540) with $x = 0$, we obtain

$$\varepsilon = [\alpha + \alpha_0(0)]\{1 + [8b/h(0)a_0]\}^{-1}, \qquad (562)$$

which indicates that large ratios of the semi-span b to the chord $h(0)$ are needed to keep the reduction to modest levels.

[†] The rolling up of the trailing vorticity referred to under (ii) above (p. 218) should accordingly be minimized under conditions of minimum induced drag, since the downward motion eqn (537) of trailing vorticity is evenly balanced across the span. It is, however, impossible to realize this condition exactly and, in practice, all trailing vortex wakes are observed to roll up.

Above all, large values of the semi-span b are needed if the minimum induced drag (eqn (560)) for given lift is to be kept down to moderate values. It is often convenient, as noted before eqn (475), to relate forces to the wing's overall planform area S. The lift coefficient C_L and drag coefficient C_D are then defined as

$$C_L = L/(\tfrac{1}{2}\rho U^2 S), C_D = D/(\tfrac{1}{2}\rho U^2 S), \tag{563}$$

and we write

$$C_D = C_{Di} + C_{Df} \tag{564}$$

as a sum of contributions from induced drag and from 'frictional' effects (more correctly, boundary-layer effects). On this basis, eqn (560) shows that the minimum value of C_{Di} can be written

$$(C_{Di})_{min} = (\pi A)^{-1} C_L^2, \tag{565}$$

where A is the 'aspect ratio' which is defined as

$$A = (2b)^2/S. \tag{566}$$

Here, A can be regarded as the ratio of the total span $2b$ to the value $S/(2b)$ of the aerofoil chord averaged along the span, so that eqn (565) once again emphasizes the importance of adequate span for minimizing drag.

For given C_{Df}, the *minimum ratio of drag to lift* given by eqns (565) and (564) is the minimum value of

$$D/L = C_D/C_L = (\pi A)^{-1} C_L + C_{Df} C_L^{-1}. \tag{567}$$

This minimum value, which is

$$(D/L)_{min} = 2(\pi A)^{-\frac{1}{2}} C_{Df}^{\frac{1}{2}}, \tag{568}$$

is realized when

$$C_L = (\pi A C_{Df})^{\frac{1}{2}}. \tag{569}$$

These equations give indications that very small ratios of drag to lift are achievable with wings of streamlined shape (permitting low frictional drag coefficient C_{Df}) and high aspect-ratio. For example, with the readily achievable values $C_{Df} = 0.01$ and $A = 8$, the minimum ratio (eqn (568)) of drag to lift is equal to 0.04. Measurements confirm that well designed wings can realize such low ratios of drag to lift—a fact which may be regarded as a triumph for fluid mechanics. Some of its countless applications are indicated in our final section.

11.4 Wings and winglike surfaces in engineering and nature

A conventional subsonic aircraft includes a single main wing pair, capable of generating (at speeds above the takeoff speed) enough lift to support the aircraft's weight. In addition, a fuselage designed to achieve low drag incorporates most of the payload being transported by the aircraft, and connects the main wing pair to a much smaller wing pair at the rear, called the tailplane (or 'stabilizer' in the US literature). The distance of the tailplane aft of the mass centre must be large enough so that, in an increase of incidence of the whole aircraft, the extra lift on the relatively small tailplane produces a *moment* about the mass centre sufficient to reduce the rate of that increase. The tailplane's attitude (or, more generally, its configuration) can, furthermore, be altered relative to the rest of the aircraft, thus increasing or decreasing the same moment so as to 'trim' the aircraft; that is, to *adjust the angle of incidence* (and so the lift coefficient) to a new level such that, at the value of $\frac{1}{2}\rho U^2$ associated with the aircraft's current altitude and speed, its weight is balanced by the lift on (primarily) the main wing pair.

Stability with respect to yaw (turning about a vertical axis) is also required. Thus, to counteract the destabilizing yawing moment (Section 8.4) on the fuselage, a vertical winglike surface or *fin*, with symmetrical aerofoil sections, must be attached at the rear. When the aircraft is yawed to port or starboard, the fin experiences a 'lift' (that is, force at right angles to the oncoming wind) which is directed horizontally and which must exert sufficient moment about the mass centre to reduce the rate of increase of angle of yaw.

The system of flight *control* uses additional aerodynamic surfaces: ailerons hinged at the rear of the outer part of both wings, and moving in opposition so as to generate a difference in lift, thus producing a rolling moment about the aircraft's longitudinal axis; a rudder hinged at the rear of the fin to produce an additional yawing moment when required; and an elevator hinged at the rear of the tailplane to produce fine adjustments in pitching moment leading to changes in angle of incidence. Roll and yaw are closely coupled; thus, in the standard manoeuvre for turning an aircraft to starboard by applying aileron action to tilt the wing pair so that the lift force on it has a component to starboard, rudder action must simultaneously be applied if undesirable 'sideslip' is to be avoided.

Finally, the aircraft must incorporate those propulsive engines that generate the thrust needed to overcome the drag on the fuselage and both the frictional and induced drag on the wings; along with fuel tanks to supply those engines over the planned distance of flight (which must include provision for emergency diversions). The ratio of drag to lift must be kept as low as possible (at least during the main, 'cruising' portion of the flight) if this planned distance, or 'range', is to be maximized. This is because the total weight of fuel on board limits the total *useful work* which can be done (that is, the work required to oppose the total drag over the required range) and because there are limits to what proportion that weight of fuel can represent of the overall aircraft weight which must be balanced by lift. Usually, it is advantageous for the main wing pair to be convertible as between two configurations: one with very low ratio of drag to lift for the main, 'cruising' portion of the flight; and another, usually requiring the extension of 'flaps', for which the lift coefficient given in expression (563) is maximized so that the aircraft weight can still be supported at the necessarily moderate speeds of takeoff and landing.

All the theories described in this book become invalid at speeds comparable with the speed of sound, at which, indeed, conventional aerofoil sections lose their effectiveness through the appearance (especially, near the leading edge on the upper surface) of 'shock waves' with consequent boundary-layer separation. This effect, however, is postponed to significantly higher speeds through the use of *sweptback wings* (Fig. 92), since the flow around each aerofoil section (that is, section normal to the span) depends primarily upon the component of the oncoming stream in the plane of the section, which is reduced as a result of sweepback. Increased maximum speed, then, is achievable through sweepback, although with the penalty of an enhanced structure weight required to support sweptback wings of a given overall *span 2b*.

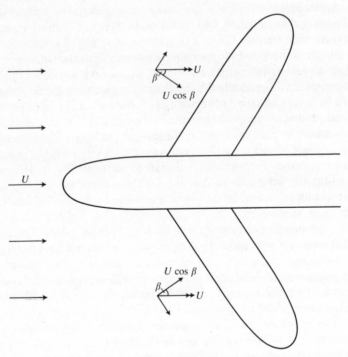

Fig. 92. Schematic illustration of sweptback wings. The oncoming airflow may be regarded as the resultant (double arrow) of:

(i) a spanwise component along the length of the wing, which is unimportant for the generation of forces; and

(ii) a component $U \cos \beta$ in the plan of the aerofoil section, where β is the angle of sweepback. It is this latter component which primarily contributes to the generation of lift.

Actually, stability is improved by sweepback. Indeed, the coupling between roll and yaw makes it advantageous that, when an aircraft's nose is yawed to starboard, flows in the planes of aerofoil sections are increased on a sweptback port wing and decreased on a sweptback starboard wing, generating a rolling moment that reduces chances of initiating a dangerous 'flat spin'. This brief remark is included mainly to indicate the great fluid-dynamic interest inherent in the theory of aircraft stability.

Not only aircraft themselves, but also their propulsive engines, incorporate many winglike surfaces. Even in cases of piston-and-cylinder engines these include *propeller blades*; while a gas-turbine engine (whether used to drive propellers or to generate jets) incorporates very much larger numbers of smaller *compressor blades* as well as *turbine blades* rotating about the engine's axis. The engine puts energy into the rotating blades of a propeller, or of a compressor, thus allowing (as we shall see) an increase of pressure in each case. Rotating turbine blades, by contrast, extract energy from the oncoming stream of fluid, rather as a windmill does; and this causes a drop in the pressure of that stream. Combustion of fuel in a gas turbine takes place in air at high pressure in the region behind the compressor; the high-temperature products of combustion then communicate much more energy to the rotating turbine blades than these need to keep the compressor moving, and the surplus drives a propeller or generates a jet.

A characteristic of all efficient systems of rotating blades is that the shape of each blade is *twisted* so that the angles of incidence of its different aerofoil sections do not vary greatly from values appropriate to high lift and low drag. Indeed, an aerofoil section at a distance r from the axis for a blade rotating at angular velocity Ω would present (see Fig. 93) a highly variable angle of incidence arctan $(\Omega r/U)$ to an axial oncoming stream of velocity U if the blade design did not incorporate twist aimed at reducing the extent of variation of that angle with r.

Different types of rotating-blade systems, on the other hand, require aerofoils with differently arranged *camber*. The reason for this can be appreciated readily if we use the frame of reference of Chapter 10, wherein the aerofoil is at rest.

With, for example, a propeller blade rotating at angular velocity Ω in an axial oncoming stream of velocity U, Fig. 93 (a) illustrates an aerofoil at distance r from the axis moving 'down the page' at velocity Ωr, while Fig. 93 (b) shows that aerofoil reduced to rest by combining the oncoming stream with a motion 'up the page' at this velocity. The combined flow is tilted upward at the above-mentioned angle, which we may write as

$$\chi, \text{ with } \tan \chi = \Omega r/U. \tag{570}$$

Now, the purpose of the propeller is to achieve thrust, which is a force from right to left in Fig. 93 (a), opposing the oncoming stream. In the steady flow of

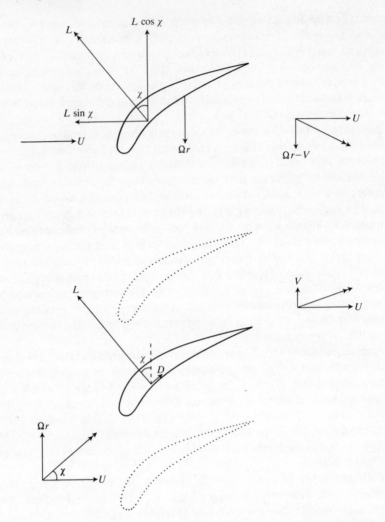

Fig. 93. Illustration of the motion of a propeller blade rotating at angular velocity Ω in an oncoming airflow of velocity U.

(a) A cross-section of the motion by the surface of a cylinder of radius r, with its axis that of the propeller (see Fig. 94), shows the blade cross-section moving down the page at velocity Ωr.

(b) Relative to the blade cross-section, the oncoming airflow is the resultant (double arrow) of a component Ωr up the page and a velocity component U from left to right. Its angle of incidence χ is given by expression (570).

Note: the lift L, shown as perpendicular to the above resultant, must be positive if the propeller blade section is to generate positive thrust $L \sin \chi$, so that the blade needs to have the type of camber shown. Dotted lines in Fig. 93 (b) represent sections of neighbouring blades by the same cylinder of radius r. Widening of the streamtube area in the passages between adjacent blades reduces the fluid speed behind them to the resultant (double arrow) of the same flow component U from left to right and a reduced flow component V up the page. This reduction implies the existence of a swirl velocity $\Omega r - V$ in the frame of reference of Fig. 93(a).

Fig. 93 (b) the lift L per unit span has a component

$$L \sin \chi \tag{571}$$

in this 'thrust' direction. Under different flight conditions different amounts of thrust will be needed but they will all be positive. Therefore, by expression (571), the aerofoil needs to be able to achieve lift coefficients that are all positive although they may cover a wide range; and this requires (Section 10.3) an aerofoil cambered as in Fig. 93.

The lift L has a further component $L \cos \chi$ tending to oppose the blade's downward motion at velocity Ωr (illustrated in Fig. 93(a)) and therefore making a contribution

$$(\Omega r) L \cos \chi \tag{572}$$

per unit span of blade to the overall *power* needed to turn the propeller. This may be compared with the contribution

$$U L \sin \chi \tag{573}$$

per unit span to the rate at which 'useful work' is done by the thrust component of expression (571) which acts on an aircraft moving through the undisturbed air at velocity U.

The circumstance that the contributions in expressions (572) and (573) are identical, by the definition (expression (570)) of χ, illustrates a most important principle: that *power conversion* from one form to another (here, from rotational to translational) would be *exactly achieved* by a system of rotating blades if the only forces acting on aerofoils were *lift* forces. In reality, of course, there must also be a drag force D per unit span (Fig. 93 (b)), which involves an *increase* $(\Omega r) D \sin \chi$ in the required power (expression (572)) but a *decrease* $U D \cos \chi$ in the rate of useful working (expression (573)). There is then an imperfect conversion of energy from rotational to translational, with some energy being 'wasted' in overcoming drag. It is nevertheless clear that blade designs capable of achieving very low *ratios* of drag to lift may be successful in achieving rather high efficiencies of energy conversion.

The total thrust T acting over the whole annular area S between the hub and the blade tips (Fig. 94) must be associated with an average increase of pressure

$$T/S \tag{574}$$

for fluid crossing that area S. This is necessary because continuity of volume flow means that the velocity component normal to S cannot change as fluid crosses S. This fluid's normal component of momentum is therefore unchanging, so that the total normal forces on it must be in balance. These forces are the reaction $(-T)$ of the propeller on the fluid and the resultant over the area S of any pressure increase; an increase whose average must therefore take the value of expression (574).

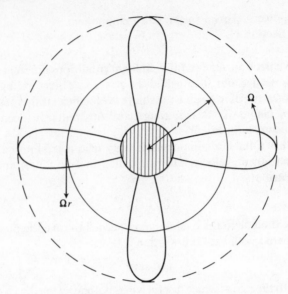

Fig. 94. Propeller viewed from the direction of the oncoming stream, illustrating that cylinder of radius r which generates the cross-sections exhibited (flattened out, of course) in Fig. 93. Across the area S between the hub (hatched) and the blade tips (broken line) the air pressure increases by an average value T/S if the propeller exerts a thrust T.

In the frame of reference of Fig. 93 (b), in which the flow around the aerofoil is steady, Bernoulli's equation (eqn (179)) requires that a pressure increase be accompanied by a decrease in $\frac{1}{2}\rho q^2$. Thus, the fluid speed q must take a reduced value downstream of the aerofoil. At first sight this may seem inconsistent with the theory of flow over lifting aerofoils (Chapter 10) which allows the velocity behind the aerofoil to return to its undisturbed value. However, if the plane of Fig. 93 (b) is regarded as a flattened-out circular cylinder of radius r, representing the surface on which the oncoming velocity components are as indicated, then the aerofoil is accompanied by sections of other blades (shown dotted), such that the aerofoil camber allows the passages between adjacent blades to widen the streamtube area in a way consistent with the reduction in q.

It is, of course, only the component Ωr of flow velocity in Fig. 93 (b) which is reduced since, as noted before, the velocity component normal to the plane of the propeller cannot change. In the frame of reference of Fig. 93 (a), this implies that the moving propeller blades generate a component of velocity 'down the page', corresponding to an azimuthal or *swirling* motion behind the whole propeller.

A complicated spiral pattern of trailing vortices separates a propeller's swirling wake from the flow outside it, and the rate of energy generation needed to continue producing that wake represents one part of the difference between the power driving the propeller and its output of 'useful work'. By analogy with Section 11.3 this part can be called *induced power*, and a classical analysis indicates that, just as a wing's induced drag is minimized (Section 11.3) when the downward flow behind it is distributed as uniformly as possible, so a propeller's induced power is minimized when the pressure increase across it varies as little as possible from its average value (expression (574)).

The rotating blades of an axial *compressor* have aerofoil sections cambered as in Fig. 93 and exploit the same mechanism to achieve an increase of pressure, although a blade's distance to its nearest neighbours (those dotted in Fig. 93) takes relatively lower values. Normally, many stages of compression are needed to achieve the overall pressure rise required. In this case, the swirling flow behind each set of *rotors* (rotating blades) is diminished by passage through a set of *stators* (stationary blades) which are cambered so as to straighten out the flow (Fig. 95). The progressive increase in pressure as each 'stage' of compression (comprising rotors followed by stators) follows the last involves a retardation which is compatible with solenoidality (Section 2.4) provided that the overall area of the channel between hub and blade tips increases continuously through a progressive diminution of the hub radius (Fig. 96).

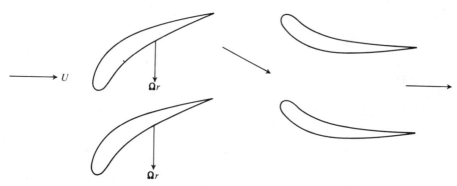

Fig. 95. Schematic illustration of the flow around neighbouring blades of an axial compressor, in a cross-section by the surface of a cylinder of radius r with its axis that of the compressor. The *rotor* blades (on the left) are rotating with angular velocity Ω. Besides generating a pressure increase, these generate swirl, which is then largely straightened out in flow through the associated system of *stator* blades (on the right).

Turbine blades must have opposite camber to the blades of Fig. 93, since the lift force needs to be directed so as to *promote* blade rotation. There is now a decrease of pressure as fluid passes the blades, corresponding (in a frame of reference in which an aerofoil section is at rest (Fig. 97)) to an increase in azimuthal velocity *above* the undisturbed value Ωr. This implies that the

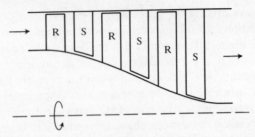

Fig. 96. Schematic illustration of the principle of the axial compressor: the retardation associated with each 'stage' of compression (rotors R followed by stators S) is accommodated by a gradual increase in the overall area of the channel between hub and blade tips. The broken line represents the axis of rotation.

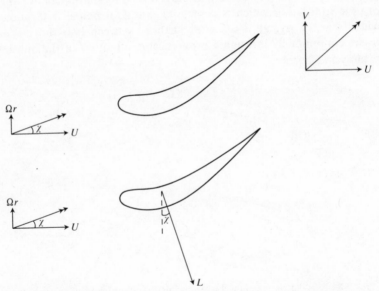

Fig. 97. Illustration of the flow, relative to the blade cross-section, around adjacent rotor blades of a gas turbine, in a cross-section by the surface of a cylinder of radius r with its axis that of the turbine. The camber must be opposite to that in Fig. 93(b) if the lift force is to be directed so as to promote blade rotation. Narrowing of the streamtube area in the passage between adjacent blades then increases the fluid speed behind them to the resultant (double arrow) of an unchanged flow component U from left to right and an increased flow component V up the page.

rotating turbine blades generate swirl in a direction opposite to their rotation. Once again, stators act to diminish this swirl before the fluid enters the next stage, while the overall area of the channel must continuously decrease as the fluid accelerates through the turbine.

By contrast, propulsive mechanisms in the animal kingdom do not involve rotary motors. Nevertheless, they often incorporate winglike surfaces and use these to achieve highly efficient propulsion. Here, we first of all describe the means of propulsion used by all of the fastest *continuously swimming* animals in the open ocean.

Biologically, it is interesting that animals of three completely distinct groups should have acquired these capabilities through a process of 'convergent evolution' which has led, for all of them, to an identical mode of propulsion through the water, based on winglike surfaces of high aspect ratio being given a characteristic mode of oscillation at the animal's posterior end. These groups (Fig. 98) include the fastest bony fishes (the scombroids, including tuna, marlin and swordfishes), the fastest cartilaginous fishes (the lamnid and isurid families of sharks) and most of the cetacean mammals (whales and dolphins). The mode of oscillation, called carangiform (after the horse-mackerel *Caranx*), necessitates a twist at each extreme of the oscillation (Fig. 99) to give backward inclination to the moving winglike surface, that is to a vertically oscillating cetacean tail, or to a fish's horizontally oscillating caudal fin.

These winglike surfaces have symmetrical aerofoil sections; evidently, the oscillatory movement requires them to achieve lift of both signs. This lift, which is necessarily acting at right angles to the direction of motion χ of the aerofoil relative to the water, exerts a thrust $L \sin \chi$ per unit span which does 'useful work' at a rate given by expression (573) per unit span. The other component $L \cos \chi$ of the lift opposes the velocity W of the oscillatory motion, whose maintenance therefore demands a power $WL \cos \chi$ which is identical with expression (573) since $W/U = \tan \chi$. Thus the conversion of energy, this time from an oscillatory to a translational mode, is once again perfect if the only forces acting are lift forces; and is highly efficient for wings achieving lifts very large compared with the sum of the frictional and induced drags. We can once more speak of 'induced power' and equate it to the rate of production of kinetic energy in a vortex wake, a wake consisting this time of a sequence of vortex rings of alternating sign.

The sharks in Fig. 98 evolved from more normal, slower groups of sharks with their characteristically asymmetrical caudal-fin arrangement: the 'heterocercal' tail dominated by a single sweptback wing (Fig. 100). Each aerofoil section of this winglike surface generates a thrusting force in the plane of the section, with components in the direction of fish motion and also in the *vertical* direction. The latter component performs the same function as the lift on an aircraft's tailplane, combining with the main lifting surface (here, the large pectoral fins) to support in a stable manner the shark's considerable excess of weight over buoyancy.[†]

[†] An excess absent for most bony fishes because they possess the gas-filled swimbladder; and unimportant for the fast, continuously mobile, animals of Fig. 98.

Fig. 98. Illustration of the lunate tail, used by all of the fastest continuously swimming animals in the open ocean. The top six animals (wahoo, tuna, marlin, louvar, sailfish and swordfish) are bony fishes of the sub-order Scombroidea, with very variously shaped front portions but with an essentially common shape of tail, which is shared by the two cartilaginous fishes (two of the fastest species of shark) shown immediately below them. All of these fishes deploy the lunate tail in a high-frequency carangiform oscillation (see Fig. 99) from side to side, and are illustrated in side view. For comparison purposes, the cetacean mammals (whale and dolphin) at the bottom of the diagram are illustrated in a view from below, since they deploy their similarly shaped lunate tail in an up-and-down carangiform oscillation.

The animals capable of sustaining continuous *flight* through the air, by generating enough lift to support body weight and enough thrust to counteract drag, fall into three groups: the insects; the birds; and the bats. None of these animals, however, adopts the separation of the thrusting organ from the main lifting surfaces which is shown in Fig. 100, or displays the man-made aircraft's clear distinction between wings generating lift and engines generating thrust. By contrast, every animal executing sustained forward flight through the air uses the same organ (the wing) to generate both.

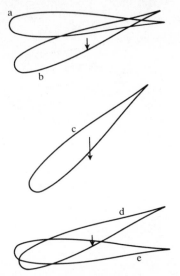

Fig. 99. Carangiform motion involves a twist at each extreme of the oscillation. Successive positions a, b, c, d, and e, assumed by a lunate tail's aerofoil cross-section at equal time-intervals, are sketched for half of the cycle of oscillation. (The other half of the cycle is a mirror image of that shown.)

Fig. 100. The blue shark, less fast than the sharks shown in Fig. 98, has a much more asymmetrical caudal-fin arrangement known as the heterocercal tail, which it oscillates from side to side to generate a force with a weight-supporting as well as a thrusting component.

The type of flapping movement needed to achieve this combination of thrust and lift is schematically indicated in Fig. 101 as a linear combination of the carangiform thrusting movement of Fig. 99 and the lift-generating capability of an aircraft wing at fixed positive angle of incidence. The wings of insects,

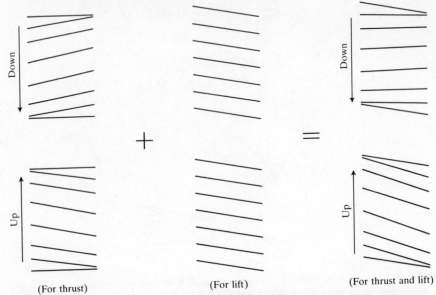

Fig. 101. Schematic illustration of the principle of flapping flight (with the animal's direction of mean motion taken towards the *left*). Thrust and lift may simultaneously be generated if the up-and-down carangiform oscillation sketched at the left of the diagram (the upstroke being depicted, for clarity, separately from the downstroke) is adapted by addition of a fixed positive angle of incidence at every phase of the oscillation, as indicated in the middle of the diagram. In the combined movement, sketched at the right of the diagram, the downstroke has the wing nearly parallel to the direction of mean motion. This movement is heavily loaded since the pressure forces generating thrust have an upward component which reinforces those responsible for generating lift. The upstroke, however, is more lightly loaded since the corresponding pressure forces generating thrust have a downward component which partly cancels those responsible for lift.

birds and bats are all able to execute the resulting combined movement with (once more) twists at each extreme of the oscillation. This involves a powerful downstroke with the wing nearly parallel to the direction of mean motion and a much more lightly loaded upstroke with the wing at a large positive angle of pitch. The diversity of wing shaping and construction is enormous, with the main emphasis placed on structure-weight minimization and on variability of wing configuration to meet the needs of different conditions of flight; these wings, indeed, do *not* possess sections of conventional aerofoil shape.

Without pursuing further that extremely interesting subject, we conclude this section with a brief reference to two methods by which animals are able to remain stationary in the air. The first of these, at which the larger birds are particularly adept, depends on finding a place where ground contours cause the local wind to be directed at a slight angle of elevation δ *above* the horizontal. At such a place an animal of weight W can remain motionless with

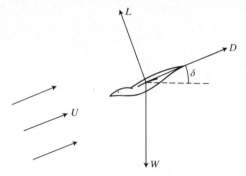

Fig. 102.   In an airflow of speed U inclined upwards at angle δ, a bird of weight W with wings
outstretched can remain stationary if its ratio of drag D to lift L is $\tan \delta$.

wings outstretched (Fig. 102) provided that the wind velocity U is sufficient to
allow those wings to develop a lift

$$L = W \cos \delta \tag{575}$$

and provided that the total drag takes the value

$$D = W \sin \delta. \tag{576}$$

The ratio of drag to lift is then

$$D/L = \tan \delta, \tag{577}$$

so that $\tan \delta$ must exceed the minimum ratio of drag to lift for the strategy to be
feasible. This illustrates yet again the advantage of designs allowing a low value
of this ratio.

The last natural phenomenon to which we refer is the aerodynamic
technique used by hummingbirds and by *most* insects[†] to remain stationary in
completely still air by making the wing motions described as 'normal'
hovering, with lift derived from reciprocating movements through the air that
are almost horizontal. These almost horizontal wing motions are a natural
development from the usual vertical motions of flapping flight, being arrived at
by turning the whole body into an almost erect position, and by greatly
accentuating the wing *twists* at each extreme of the oscillation, so that as the
wing moves back and forth the same leading edge is always leading. Still more
precisely, the tip of the wing moves in a figure-of-eight pattern (Fig. 103) such

---

[†] Various other hovering techniques (sometimes referred to as 'exceptional') are used by certain
groups of insects, including dragonflies and also all *very small* insects (for which the Reynolds
number is too low for the 'normal' hovering technique to be effective).

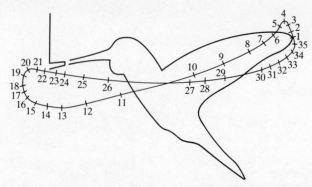

Fig. 103. Figure-of-eight pattern of the path of the wing tip during the sustained hovering of a hummingbird. The points represent successive positions of the wing tip at equal time intervals.

that, especially during the heavily loaded 'downstroke' (defined here as the stroke during which the wing tip moves from a *dorsal* to a *ventral* position), the wing path is directed at an angle below the horizontal so that, as in Fig. 102, the resultant of the lift and drag on the wing has a strong vertically upward component.

The purpose of this concluding section has been to indicate wide areas of application of knowledge regarding wings and winglike surfaces able to achieve low ratios of drag to lift. All of them are areas where detailed theoretical studies of the types set out in this book, concerned with irrotational flows and boundary layers and shed vorticity, have been, and may still be, most fruitfully applied.

Exercises

(Note: the *first* of the two figures attached to each exercise represents the number of chapter on which the exercise is set.)

1.1. Explain why the speed with which water at atmospheric pressure will rush into a vacuum (without change of level) is about $14 \, \mathrm{m \, s^{-1}}$.

1.2. Water flows steadily along a horizontal pipe, out of which two glass tubes stretch vertically upwards. Each of these tubes is open at the top, and is partly full of water at rest in contact with the flowing water. The cross-section of the pipe is $3 \times 10^{-3} \, \mathrm{m^2}$ at the first tube's attachment point and $2 \times 10^{-3} \, \mathrm{m^2}$ at that of the second tube (farther downstream). Show how an observation, that the water level in the first tube is 0.1 m above the water level in the second, may be used to infer that the volume flow through the pipe is about $3.7 \times 10^{-3} \, \mathrm{m^3 \, s^{-1}}$.

1.3. A cylindrical vessel of large diameter has a waste pipe of length 0.1 m and cross-section $4 \times 10^{-5} \, \mathrm{m^2}$ protruding vertically downwards from the centre of its horizontal base. Water enters the vessel at the rate of $10^{-4} \, \mathrm{m^3 \, s^{-1}}$. Show that, in the steady state when the rate of outflow from the waste pipe has reached this value, the vessel is filled to a depth of about 0.22 m.

1.4. Water from a firehose of internal cross-section $5 \times 10^{-3} \, \mathrm{m^2}$ emerges horizontally through a nozzle of internal cross-section $10^{-3} \, \mathrm{m^2}$ at a speed of $20 \, \mathrm{m \, s^{-1}}$. Show that the fireman needs to exert a force of about 640 N to hold on to the hose.

1.5. A jet of water of cross-section $2 \times 10^{-5} \, \mathrm{m^2}$ and speed $5 \, \mathrm{m \, s^{-1}}$ is deflected through 60° by partly inserting a finger. Show that, in a steady state, the force which must be exerted on the finger to prevent it being sucked further into the jet is about 0.5 N.

1.6. Water flows along a horizontal pipe ABC out of a large reservoir, the pipe having a smoothly shaped entry at A and being at a distance h below the surface level in the reservoir. The part AB of the pipe has uniform cross-section S, but then there is an abrupt widening and the part BC has cross-section $2S$. Calculate the speed of the jet issuing from the pipe into the atmosphere at C.

1.7. The resistance coefficient k for a pipe is sometimes defined as the difference of total head between the two ends of the pipe divided by the *square* of the volume flow through the pipe. This definition makes possible a calculation of the resistance for a network of pipes, similar to that used in electric current theory (where, however, the potential difference between the two ends of a wire is *directly* proportional to the current in the wire). Show, for example, that for two pipes of resistance coefficients k_1 and k_2 the effective resistance coefficient k for a system consisting of the two pipes 'in parallel' is

$$k = k_1 k_2 (\sqrt{k_1} + \sqrt{k_2})^{-2}.$$

2.1.   If the position vector $\mathbf{r} = (x, y, z)$ has magnitude r, and $f(r)$ is any differentiable function, show that
$$\text{grad } f(r) = f'(r)\, r^{-1}\, \mathbf{r}.$$

2.2.   If $p = (x^2 + y^2 + z^2)^{\frac{1}{2}}$ and $q = xy/z^2$, calculate grad p and grad q. Show that the scalar product $(\text{grad } p) \cdot (\text{grad } q)$ is zero. Can you think of any way by which this result could have been predicted in advance without separately calculating grad p and grad q?

2.3.   The fluid velocity at (x, y, z) is $(cx, cy, -2cz)$, where c is a constant. Prove that the velocity field is solenoidal, and find an equation for the shape of the streamline through the point (x_0, y_0, z_0).

2.4.   The fluid velocity at (x, y, z) is $(cxs^{-3}, cys^{-3}, 0)$, where $s^2 = x^2 + y^2$. Prove that the velocity field is solenoidal, and find an equation for the shape of the streamline through the point (x_0, y_0, z_0).

2.5.   Prove that the integral over a closed surface of the length of the perpendicular from the origin on to a plane tangential to the surface is three times the volume enclosed by the surface.

2.6.   A pressure which is proportional to the square of the distance from the origin O acts over the surface of a homogeneous body. Prove that the resultant force is in the direction GO, where G is the centre of gravity of the body.

2.7.   Verify the Divergence Theorem by evaluating
$$\int_{L} \text{div } \mathbf{u}\; \mathrm{d}V \text{ and } \int_{\partial L} \mathbf{u} \cdot \mathbf{n}\; \mathrm{d}S,$$
where L is the sphere $x^2 + y^2 + z^2 \leqslant a^2$ and
$$\mathbf{u} = (2xy^2 + 2xz^2, x^2 y, x^2 z).$$

3.1.   In a certain fluid motion, the velocity field is $\mathbf{u} = (ax, by, cz)$ where a, b and c are constants. Show that the fluid density ρ takes the value $\rho = \rho_0 \exp[(a+b+c)t]$ at time t, given that $\rho = \rho_0$ throughout the fluid when $t = 0$.

3.2.   The outlet from a very large reservoir consists of a straight horizontal pipe of length l whose cross-section S takes the value $S = S_0 (1 + xl^{-1})^{-1}$, where x is the distance from the reservoir. Until the time $t = 0$ the outlet is blocked but then it is opened. If the volume flow through the pipe for $t > 0$ takes the form $Q(t)$ show that the fluid acceleration may be written
$$(1 + xl^{-1})\{S_0^{-1}\dot{Q}(t) + l^{-1}S_0^{-2}[Q(t)]^2\}.$$

Neglecting any variation in the height h of the water level in the reservoir above the level of the pipe (whose resistance should also be neglected), derive the differential equation
$$l\dot{Q}(t) + S_0^{-1}[Q(t)]^2 = \tfrac{2}{3}ghS_0$$
and use it to determine the function $Q(t)$.

3.3.   Within a fluid, a solid sphere of radius a moves in a circle of radius c so that at time t the coordinates of its centre are $(0, c\cos \omega t, c\sin \omega t)$. After writing the equation of the sphere's surface in the form of eqn (75), deduce that the fluid velocity adjacent to it satisfies the boundary condition

$$ux + v(y - c \cos \omega t) + w(z - c \sin \omega t) = \omega c(z \cos \omega t - y \sin \omega t).$$

3.4.   From a long horizontal slit in the vertical wall of a large reservoir, there emerges a jet in the form of a horizontal slab of water of depth h moving at a high velocity U. While still horizontal (because it has not travelled far enough to be significantly bent downwards by gravity) the jet strikes a plank held at rest with the plank's length parallel to the slit and with the flat surface of the plank inclined at an angle α to the horizontal, upwards and away from the reservoir. Part of the jet turns through an angle α to run upwards along the surface of the plank, and the remainder turns through an angle $\pi - \alpha$ to run downwards along the same surface. Explain why the velocity of both parts must be U (if, as before, it is possible to neglect effects of gravity and also of any frictional resistance at the plank surface). From the fact that the water pressures must exert a force on the plank that acts perpendicularly to its surface, and is equal and opposite to the reaction of the plank on the water, show that this force must have the magnitude $\rho U^2 h \sin \alpha$ per unit horizontal length of plank; and, also, that the thicknesses of the fluid motions up and down the plank are $\frac{1}{2} h (1 + \cos \alpha)$ and $\frac{1}{2} h (1 - \cos \alpha)$, respectively.

4.1.   If $\mathbf{u} = (2x^2 y, xz^2 - y^3, xyz)$, calculate div $\mathbf{u}$, curl $\mathbf{u}$ and div curl $\mathbf{u}$.

4.2.   If $\mathbf{u}$ and $\mathbf{v}$ are any two vector fields, show that the divergence of their vector product takes the form

$$\text{div } (\mathbf{u} \times \mathbf{v}) = \mathbf{v} \cdot \text{curl } \mathbf{u} - \mathbf{u} \cdot \text{curl } \mathbf{v}.$$

By applying the Divergence Theorem to $\mathbf{u} \times \mathbf{v}$ where $\mathbf{v}$ is any *constant* vector, show that

$$\int_L \text{curl } \mathbf{u} \ dV = \int_{\partial L} \mathbf{n} \times \mathbf{u} \ dS.$$

4.3.   Each particle of water in an open vessel is moving in a circular path about a vertical axis with a speed $q(s)$ depending only on the radius s of the path. Prove that the vorticity is a vertically directed vector with magnitude $q'(s) + s^{-1} q(s)$.
If the magnitude of the vorticity takes a constant value ω_0 for $s < a$ and is zero for $s > a$, determine the value of $q(s)$ in both regions. Also, find the pressure from the momentum equation (taking gravity into account) and deduce the shape of the free surface, showing it to be lower by a distance $(\omega_0^2 a^2 / 4g)$ at the centre of the vortex than it is in the undisturbed fluid.

4.4.   In the velocity field $(cy^2 + cz^2, 2cxy, - 2cxz)$, where c is constant, show that the density of each fluid element is constant. Show that the vortexlines are all parallel straight lines, and that along any very thin vortextube the magnitude of the vorticity takes a constant value. At what points are the principal axes of rate of strain parallel to the coordinate axes?

4.5.   In cylindrical coordinates (s, θ, z) defined so that $x = s \cos \theta$, $y = s \sin \theta$, the velocity components in the radial direction (s increasing) and in the azimuthal direction (θ increasing) are written u and v, being related to the ordinary x and y components of velocity u and v by equations

$$\text{u} = u \cos \theta + v \sin \theta, \quad \text{v} = v \cos \theta - u \sin \theta.$$

The component of velocity in the axial direction (z increasing) is, of course, simply w. A box-shaped particle of fluid is specified by the coordinates being limited to intervals $(s, s + ds)$, $(\theta, \theta + d\theta)$, and $(z, z + dz)$. Applying Stokes's theorem to different faces of the box-shaped particle, show that the components of vorticity in the radial, azimuthal, and axial directions are

$$\left(\frac{1}{s} \frac{\partial w}{\partial \theta} - \frac{\partial v}{\partial z}, \frac{\partial u}{\partial z} - \frac{\partial w}{\partial s}, \frac{1}{s} \frac{\partial sv}{\partial s} - \frac{1}{s} \frac{\partial u}{\partial \theta} \right).$$

4.6.   Verify Stokes's theorem (eqn (134)) for a vector field with components

$$u = 0, \quad v = z \sin \theta, \quad w = s \cos \theta$$

in the notation of Exercise 4.5, when the surface S is specified in cylindrical polar coordinates as

$$s = a, \quad 0 \leqslant \theta \leqslant \pi, \quad 0 \leqslant z \leqslant b.$$

5.1.   Show that the shearing motion of Fig. 21 satisfies the Euler equations of motion everywhere (with what distribution of pressure?). Show also that, in this motion, any closed curve moving with the fluid in a plane $z = $ constant encloses always the same area. Why can we say that these two statements taken together constitute a specific verification of Kelvin's theorem?

5.2.   For any fluid motion it is possible to consider that particle of fluid whose position vector was $\mathbf{r}$ at time $t = 0$ and to ask where the particle so defined has moved to at a particular later time t. If its position vector takes the value $\mathbf{R}$ at time t then we may regard $\mathbf{R}$ as a vector field in relation to its variation with the initial position $\mathbf{r}$ of the particle. With $\mathbf{R}$ so regarded as a vector field, and $\boldsymbol{\omega}$ taken as the vorticity of the particle at time $t = 0$, investigate the expression $\boldsymbol{\omega} \cdot \nabla \mathbf{R}$. Show how Helmholtz's theorem implies that this expression represents the particle's vorticity at the later time t.

5.3.   Check by means of the following example that, even in an irrotational motion, a small particle of fluid that is *not* spherical is able to rotate. The irrotational motion (eqn (155)) outside a line vortex allows a thin needle-shaped particle of fluid (which initially stretches radially outward at distances between r and $r + dr$ from the vortex) to move in such a way that when the nearer end has traversed an angular distance α, the farther end has traversed an angular distance $\alpha(1 - 2r^{-1} dr)$. Deduce that the rod has rotated through an angle $\alpha - \arctan (2\alpha)$, which is negative for $0 < \alpha < 1.17$ (measured in radians) and positive for $\alpha > 1.17$.

5.4.   Calculate the distribution of temperature given in Fig. 31, which results from diffusion of heat with the diffusivity in eqn (161), after a cold body is immersed in hot liquid. A thin slice of the body, of unit area, at distances from the boundary between z and $z + dz$, is gaining heat transferred across unit area at the rate $-k\partial T/\partial z$ and losing it at a rate equal to the same quantity with z replaced by $z + dz$, thus giving a net rate of gain $k \, (\partial^2 T/\partial z^2) \, dz$. But the thin slice has mass ρdz so that the rate of increase of heat can be written $c \, (\rho dz) \, \partial T/\partial t$ where c is the specific heat. Deduce the differential equation for the temperature T and look for a solution in which T has the functional form

$$T = f(\eta) \text{ with } \eta = z \, (\kappa t)^{-\frac{1}{2}}.$$

Show that

$$f'(\eta) = C \exp \left(-\tfrac{1}{4} \eta^2 \right),$$

where C is a constant, and deduce the solution which satisfies boundary conditions

$$f = T_h \text{ at } \eta = 0 \text{ and } f \to T_c \text{ as } \eta \to \infty$$

in terms of the indefinite integral of the above exponential.

5.5.   A particularly simple boundary layer can be calculated by a method closely similar to that of Exercise 5.4, as Rayleigh showed. A very large expanse of fluid (in the region $z > 0$) is bounded by a solid plane boundary ($z = 0$). The fluid is at rest when the boundary suddenly starts to move with velocity $(U, 0, 0)$ in its own plane. Show, for a thin slice of fluid of unit area at distances from the solid boundary between z and $z + dz$, that the net rate of gain of x component of momentum is $\mu(\partial^2 u / \partial z^2)\, dz$, and deduce a differential equation

$$\partial u / \partial t = v \partial^2 u / \partial z^2$$

for the velocity u. Give a clear explanation of what causes this motion to be unusual in that: (i) exceptionally, only the single term on the left-hand side of the above equation represents the fluid acceleration; and (ii) no pressure gradient term is present. (Note: calculations of boundary layers in general are complicated, because they lack *both* these simplifying features.) Finally, adapt your solution to Exercise 5.4 so as to obtain the form of u satisfying the above equation together with boundary conditions appropriate to the present problem.

6.1.   For the vector field $\mathbf{u}$ defined as

$$\mathbf{u} = (z - 2xr^{-1},\; 2y - 3z - 2yr^{-1},\; x - 3y - 2zr^{-1}),$$

where $r = (x^2 + y^2 + z^2)^{\frac{1}{2}}$, prove that curl $\mathbf{u} = 0$ and find ϕ such that $u = $ grad ϕ.

6.2.   Prove that the angular momentum about its centre of a liquid spherical volume of any radius a is zero for irrotational motion.

6.3.   Prove that, for *steady* irrotational flow of a compressible fluid of nonuniform density ρ, the velocity potential ϕ satisfies an equation of continuity

$$\nabla \cdot (\rho \nabla \phi) = 0.$$

Deduce that a steady irrotational flow of a compressible fluid in a finite simply connected region with its boundaries at rest has zero kinetic energy, so that the fluid itself is at rest. Note: for incompressible fluid, this result is proved in Section 6.2 to be true whether or not the flow is steady. For compressible fluid, however, non-zero irrotational motions (known as sound waves!) are possible if the flow is not restricted to being steady. Why is your proof invalid in the unsteady case?

6.4.   The analysis of Section 6.3 for a straight line vortex near a plane solid boundary, leading to the conclusion that the vortex moves parallel to the plane, can be extended to allow consideration of the case when the vortex, having arrived at a 90° corner with a second wall, turns the corner to follow the second wall. Suppose that fluid fills the region $x > 0$, $y > 0$ which is bounded by walls $x = 0$ and $y = 0$ forming such a corner. Show that a vortex of strength K at the point (k, h) generates a flow associated with: (i) the vortex and its image $(k, -h)$ in the boundary $y = 0$; together with (ii) the images of both of these in the boundary $x = 0$. Show that the motion of the vortex itself is caused by the combined flow field of three image vortices (at $(k, -h)$, $(-k, h)$, and

$(-k, -h)$) so that the coordinates of the vortex change in accordance with the equations

$$dk/dt = ck^2h^{-1}(k^2+h^2)^{-1}, dh/dt = -ch^2k^{-1}(k^2+h^2)^{-1},$$

where $c = K/(4\pi)$. Determine the path of the vortex given that $k = k_0$ when h is very large.

7.1.   If $r = (x^2 + y^2 + z^2)^{\frac{1}{2}}$, find $\nabla^2(r^n)$ and prove that it is zero only when $n = 0$ or $n = -1$.

7.2.   At the apex of a large fluid-filled circular cone of semi-angle α, a volume J of fluid is sucked out of the interior of the cone during each second. Find the velocity potential of the irrotational flow inside the cone.

7.3.   A liquid contains an empty spherical cavity of radius $a(t)$ and internal pressure zero. The pressure in the liquid far from the cavity is P. Initially the liquid is at rest and the cavity radius is a_0. In the analysis of Section 7.2 (neglecting gravity and surface tension) show that until the radius $a(t)$ has reached the value $4^{-\frac{1}{3}}a_0$ the greatest pressure in the fluid is P, a maximum attained far from the cavity, but that for lower values of a/a_0 the greatest pressure $p_{\max}$ is found where

$$r/a = [(a_0^3 - a^3)/(\tfrac{1}{4}a_0^3 - a^3)]^{\frac{1}{3}}.$$

Evaluate $p_{\max}/P$ as a function of a/a_0.

7.4.   A source and sink each of strength J and at a distance $2h$ apart in the direction of a stream of velocity U give the irrotational flow around a Rankine ovoid of length $2l$. Show that

$$J = \pi U (l^2 - h^2)^2 l^{-1}h^{-1}.$$

7.5.   The irrotational flow with velocity $(0, 0, U)$ past the finite body of revolution

$$s = R(z) \qquad (-l \leqslant z \leqslant l),$$

where $s = (x^2 + y^2)^{\frac{1}{2}}$, may be represented by a continuous distribution of sources along the axis $s = 0$, with strength $f(z)$ per unit length. Show that the function $f(z)$ must satisfy the integral equation

$$2\pi U R^2(z) = \int_{-l}^{l} f(Z)(z-Z)[(z-Z)^2 + R^2(z)]^{-\frac{1}{2}}dZ.$$

If, as may be justifiable for a *slender* body, $(z-Z)[(z-Z)^2 + R^2(z)]^{-\frac{1}{2}}$ is approximated by $+1$ for $z > Z$ and by -1 for $z < Z$, show from the integral equation that

$$f(z) = U d[\pi R^2(z)]/dz.$$

Can you see a physical interpretation for this approximate value of the source strength?

7.6.   A sphere of radius a is placed in a steady stream of velocity U. For irrotational flow, how far upstream of the sphere's centre is the velocity reduced to $0.95\,U$? Show that excess pressures on (i) the half of the sphere which faces the stream and on (ii) the half facing away from the stream each have negative resultants of magnitude

$$\frac{1}{16}\pi\rho U^2 a^2$$

tending to pull the two hemispheres apart.

7.7.   Not only the velocity $-\mathbf{U}(t)$ of the centre of a solid sphere but also its *radius* $a(t)$ are functions of the time t. The sphere is immersed in fluid of density ρ, which is in irrotational motion with its velocity $\mathbf{u}$ tending to zero at distances far from the body. Show that the velocity of the fluid at a point whose position vector relative to the centre of the sphere is $\mathbf{x}$ takes the value

$$\mathbf{u} = \tfrac{1}{2}a^3r^{-3}\{[2\dot{a}a^{-1} - 3(\mathbf{U}\cdot\mathbf{x})r^{-2}]\mathbf{x} + \mathbf{U}\},$$

where $r = |\mathbf{x}| > a$. If $\mathbf{U}$ is a constant vector, and if $\dot{a} = |\mathbf{U}|$, show that the excess pressure p_e takes the value $\tfrac{1}{2}\rho U^2$ at that point of the sphere which remains at rest.

8.1.   Let ϕ_n stand for any particular solution of Laplace's equation which is a homogeneous polynomial of degree n in x, y, and z. (For example, ϕ_2 might be $x^2 + y^2 - 2z^2$.) Putting $S_n = \phi_n r^{-n}$ show that an equation $\mathbf{r}\cdot\nabla S_n = 0$ expresses the fact that S_n is a homogeneous function of degree *zero*. Prove that $S_n r^{-n-1}$ is a solution of Laplace's equation and show that the potentials of a source and of a dipole take this form for $n = 0$ and $n = 1$ respectively. By an appropriate use of Green's theorem, show that for any S_m and S_n with $m \neq n$ the integral

$$\int_{\Sigma_r} S_m S_n \, dS$$

over any sphere Σ_r of radius r and centre the origin takes the value zero.

8.2.   Show that

$$[x^2 + y^2 + (z-h)^2]^{-\frac{1}{2}} = (r^2 - 2zh + h^2)^{-\frac{1}{2}}$$

is a solution of Laplace's equation, and infer that each term in its expansion (for large r) in descending powers of r,

$$(r^2 - 2zh + h^2)^{-\frac{1}{2}} = \sum_{n=0}^{\infty} h^n P_n r^{-n-1},$$

is a solution of Laplace's equation. In other words, P_n is a special case of the general S_n defined in Exeircse 8.1, its main special feature being that it is a function $P_n(\theta)$ of the single variable θ defined by the equation

$$z/r = \cos\theta.$$

Carry out the above expansion in descending powers of r, showing that

$$P_2 = \tfrac{1}{2}(3\cos^2\theta - 1)$$

and determining the value of P_3. Use these results for the flow with velocity $(0,0,U)$ around the Rankine ovoid of Fig. 48 so as to find the next term in the expansion of the potential for large r after the leading terms

$$Uz - (hJ/2\pi)zr^{-3}$$

associated with the undisturbed flow and with the strength (eqn (322)) of the equivalent dipole.

8.3. Calculations have shown that, when a circular disc of radius a is moving normal to its plane at speed U through otherwise undisturbed fluid which is in irrotational motion, the velocity potential at the surface of the disc takes the values

$$\phi = \pm (2U/\pi)(a^2 - s^2)^{\frac{1}{2}},$$

where s represents distance from the disc's centre and the two signs correspond to the two faces of the disc. Show that the total kinetic energy of the fluid is $\frac{4}{3}\rho a^3 U^2$ and determine the far-field dipole strength. Consider the instantaneous motion generated when a heavy body with a flat circular base falls with speed U on to the horizontal free surface of a large expanse of water at rest. Taking into account the boundary condition at the free surface (Section 3.4), show that the upward impulse on the body resulting from the impact is $\frac{4}{3}\rho a^3 U$.

8.4. Pursuing further the analysis (Section 8.4) of any 'slender' body of revolution $s = R(z)$ (such as the Rankine ovoid) which is moving at a uniform velocity $(U,0,0)$ at right angles to its axis (the z axis) in otherwise undisturbed fluid, obtain an approximate estimate of the velocity potential

$$\phi = -\frac{s\cos\psi}{4\pi}\int_{-l}^{l}\frac{g(Z)\mathrm{d}Z}{[s^2 + (z-Z)^2]^{\frac{3}{2}}}$$

which is associated with a distribution of dipoles (eqn (326)) stretching along the length $-l \leqslant Z \leqslant l$ of the body. For a given point (s,z) on the body surface, use the fact that

$$|g(Z) - [g(z) + (Z-z)g'(z)]| \leqslant \tfrac{1}{2}M(Z-z)^2,$$

where M is the upper bound of $|g''(z)|$, to estimate the error in the expression

$$\phi = -\frac{g(z)\cos\psi}{2\pi s}$$

(see eqn (328)) from which is derived the approximate form

$$g(z) = 2\pi U[R(z)]^2$$

for the strength $g(z)$ of the transverse dipoles.

9.1. The function $Z(z) = z^2$ maps the domain D which consists of the *upper half z plane* defined by

$$z = re^{i\theta}\,(0 < r, \quad 0 < \theta < \pi)$$

onto a domain E. Describe E and ∂E carefully. Show that the simple uniform flow in E represented by the complex potential $f(Z) = UZ$ is mapped onto a flow in D with a stagnation point on the boundary, where the dividing streamline impinges on the stagnation point at right angles to the boundary, and show that the conditions for an irrotational flow to be physically realizable are met. (Note: this flow in D is typical of the *local* flow very close to any stagnation point on a smooth boundary.) Show also that the uniform flow in D represented by the complex potential $f(z) = Uz$ is mapped onto an irrotational flow in E that does not meet the conditions for the flow to be physically realizable. In this last flow, what happens to the fluid velocity as $Z \to 0$?

9.2.   In the z plane, the domain D is defined by

$$z = x + iy, \qquad 0 < y < h,$$

so that its boundary ∂D consists of two parallel lines. Show that the function $Z(z) = \exp(\pi z/h)$ maps D onto a domain E in the Z plane, and describe E and ∂E carefully. Out of *two* simple uniform flows in D, represented by the complex potentials

(a) $f(Z) = +UZ$ and   (b) $f(Z) = -UZ$,

show that the mapping converts one of them, but not the other, into an irrotational flow in D which meets the conditions for the flow to be physically realizable. What names are given to these two flows?

9.3.   The function $Z(z)$ conformally maps the domain D in the z plane onto the domain E in the Z plane. Show that the correspondence between any complex potential $f(z)$ of a two-dimensional irrotational flow in D and the complex potential of the same flow mapped onto E possesses the following properties.
   (i)   If the flow in D includes a *line source* of strength j per unit span at an interior point $z = z_0$, alongside other motions without singularities at $z = z_0$, so that from eqn (373) the function

$$f'(z) - (j/2\pi)(z - z_0)^{-1}$$

must be analytic at $z = z_0$, show that the flow in E includes a line source of the *same* strength j per unit span at the corresponding point $Z = Z_0$. Why is this conclusion not necessarily valid if z_0 is on the boundary ∂D?
   (ii)   Consider similarly the case when the flow in D includes a line vortex of strength K at $z = z_0$, so that from eqn (375) the function

$$f'(z) + (iK/2\pi)(z - z_0)^{-1}$$

must be analytic and must tend to a finite limit w_0 at that point. Show first that the line vortex moves in the domain D of the (x, y) plane with a velocity (u_0, v_0) given by the complex velocity $u_0 - iv_0 = w_0$. For the corresponding flow in E, show that it includes a line vortex of the same strength K at $Z = Z_0$, and then prove that the complex velocity W_0 with which this vortex moves is related to w_0 by the equation

$$w_0 = W_0 Z'(z_0) - iK Z''(z_0)[4\pi Z'(z_0)]^{-1}.$$

Verify that, when $Z(z)$ is the mapping of Exercise 9.2, the velocity w_0 with which a vortex of strength K placed at the point $z_0 = iy_0$ with $0 < y_0 < h$ moves as a result of the presence of the two stationary plane boundaries $y = 0$ and $y = h$ is

$$w_0 = (K/4h) \cot(\pi y_0/h).$$

9.4.   In the z plane, the domain D_+ is defined by

$$z = x + iy, \qquad 0 < x, \quad 0 < y < h,$$

so that it is the right-hand half of the domain D of Exercise 9.2. A line source of strength j per unit span is placed at the interior point $z = x_0 + iy_0$ of D_+. Define a flow in D symmetrical about $x = 0$, associated with this source and an image source, which satisfies the boundary conditions of zero normal velocity on ∂D_+. Show that its complex potential is

$$f(z) = (j/2\pi)\{[e^{\pi z/h} - e^{\pi(x_0 + iy_0)/h}]^{-1} + [e^{\pi z/h} - e^{\pi(-x_0 + iy_0)/h}]^{-1}$$
$$+ [e^{\pi z/h} - e^{\pi(x_0 - iy_0)/h}]^{-1} + [e^{\pi z/h} - e^{\pi(-x_0 - iy_0)/h}]^{-1}\}.$$

10.1.   For the irrotational flow (eqn (422)) with velocity U around a circular cylinder of radius a, consider the exceptional case when the circulation K around the cylinder takes the particular value $4\pi a U$ for which the two stagnation points (expression (426)) coincide. Show that in this exceptional case (produced by the fusion of two 'ordinary' stagnation points) the dividing streamline approaches the stagnation point $z = -ia$ at an angle of $60°$ to the surface and also leaves it again at such a $60°$ angle. Sketch the shape of the streamlines very close to the dividing streamline.

10.2.   Show that, in the mapping $Z = z + a^2 z^{-1}$ used extensively in Chapters 9 and 10, the derivative of the mapping has the limiting property

$$(Z + 2a)(dz/dZ)^2 \to -\tfrac{1}{4}a \text{ as } Z \to -2a.$$

Let $f(z)$ be any complex potential in the z domain D. Show that the corresponding complex potential in the Z domain E has a complex velocity $w = df/dZ$ such that the function w^2 possesses a *pole* at $Z = -2a$, with residue

$$-\tfrac{1}{4}a[f'(-a)]^2.$$

This pole is important for any sort of analysis in which we approximate the flow around an aerofoil by flow around a flat plate with a sharp leading edge (regarded as crudely representing the real blunt edge). Superficially, the result in expression (473) appears surprising for a flat plate stretching from $Z = -2a$ to $Z = 2a$. The pressure differences across the plate would be expected to have a resultant $(0, G, 0)$ strictly in the Y direction, and their integral comes in fact to $G = L\cos\alpha$. The idea that F should necessarily be zero is incorrect, however. The reader is invited to calculate the contribution to the integral (eqn (463)) from a very small contour encircling the pole of w^2 at $Z = -2a$, obtaining the value

$$F = \tfrac{1}{4}\pi\rho a[f'(-a)]^2$$

for the X component of the force per unit span. For flow about a flat plate with velocity U at angle of incidence α, and with circulation related to angle of incidence as in (eqn (456)), show that

$$F = -4\pi\rho a U^2 \sin^2\alpha$$

in agreement with expression (473). This force in the negative X direction is often called the *leading-edge suction*. Note that when the sharp leading edge is replaced by the blunt leading edge of a very thin Zhukovski aerofoil (or ellipse) the contribution to the integral (eqn (463)) for the force per unit span from the immediate neighbourhood of the leading edge is a force which tends to the above limiting value as the edge becomes sharp.

10.3.   The cambered Zhukovski aerofoil is obtained by using the mapping $Z = z + a^2 z^{-1}$ on the circle defined by expressions (478) and (479). Show that its limit as $\delta \to 0$ is a circular arc passing through the points $Z = +2a$ and $Z = -2a$. (Note: the easiest method is to show that for any point Z on the aerofoil the angle between the lines joining that point to $Z = +2a$ and to $Z = -2a$ takes the constant value $2\arctan(a/\varepsilon)$.)

11.1. Show that the potential of the rectangular line vortex in Fig. 84 may be written

$$\frac{K}{4\pi}\left[\arctan\left\{\frac{y[(x-a)^2+y^2+z^2]^{\frac{1}{2}}}{(x-a)z}\right\}-\arctan\left\{\frac{y[(x+a)^2+y^2+z^2]^{\frac{1}{2}}}{(x+a)z}\right\}\right.$$

$$\left.-\arctan\left\{\frac{y[(x-a)^2+y^2+(z-c)^2]^{\frac{1}{2}}}{(x-a)(z-c)}\right\}+\arctan\left\{\frac{y[(x+a)^2+y^2+(z-c)^2]^{\frac{1}{2}}}{(x+a)(z-c)}\right\}\right].$$

Verify that for very small values of a and c this can be written as the potential of a dipole of strength $(0,-2acK,0)$.

11.2. For three-dimensional flow around a wing show that such a large departure from the optimum distribution of circulation (eqn (561)) along the span $-b \leqslant x \leqslant b$ as the distribution

$$K = K_{max}(1-x^2b^{-2})^{\frac{3}{2}}$$

causes the induced drag D_i for a given lift L to take a value greater than its theoretical minimum (eqn (560)) by a factor $\frac{4}{3}$.

11.3. There are certain engineering systems in which just a single set of turbine blades, rotating always in the same direction in their own plane, is required to be able to extract energy from a motion of fluid across that plane either from the front or from the rear. Consider the advantages (and any disadvantages) of the 'Wells Turbine' design for this purpose, that is of a design in which the sections of all blades are symmetrical aerofoils (symmetrical about that plane).

Index